REAL QUATERNIONIC CALCULUS HANDBOOK

Real Quaternionic Calculus Handbook

Editor

Qing Jun Hou

scitus academics

Real Quaternionic Calculus Handbook
Edited by **Qing Jun Hou**

ISBN: 978-1-68117-567-6
Library of Congress Control Number: 2016942998

SCITUS Academics LLC,
www.scitusacademics.com
Box No. 4766, 616 Corporate Way,
Suite 2, Valley Cottage,
NY 10989

Printed in United States of America on Acid Free Paper

PREFACE

A quaternion is a four-element vector that can be used to encode any rotation in a 3D coordinate system. Technically, a quaternion is composed of one real element and three complex elements, and it can be used for much more than rotations. Quaternions provide an alternative measurement technique that does not suffer from gimbal lock. Quaternions are less intuitive than Euler Angles and the math can be a little more complicated. Quaternions find uses in both theoretical and applied mathematics, in particular for calculations involving three-dimensional rotations such as in three-dimensional computer graphics, computer vision and crystallographic texture analysis. In practical applications, they can be used alongside other methods, such as Euler angles and rotation matrices, or as an alternative to them, depending on the application. In modern mathematical language, quaternions form a four-dimensional associative normed division algebra over the real numbers, and therefore also a domain.

Real Quaternionic Calculus Handbook is a comprehensive guide aims to describe phenomena in special relativity, electrodynamics, spin etc., it has developed into a body of material that relates with many branches of mathematics, such as complex analysis, harmonic analysis, differential geometry, and differential equations.

TABLE OF CONTENTS

1

ENABLING QUATERNION DERIVATIVES: THE GENERALIZED HR CALCULUS

Dongpo Xu1, Cyrus Jahanchahi2, Clive C. Took3 and Danilo P. Mandic2

1School of Mathematics and Statistics, Northeast Normal University, Changchun 130024, People's Republic of China

2Department of Electrical and Electronic Engineering, Imperial College London,London SW7 2AZ, UK

3Department of Computing, University of Surrey, Guildford GU2 7XH, UK

ABSTRACT

Quaternion derivatives exist only for a very restricted class of analytic (regular) functions; however, in many applications, functions of interest are real-valued and hence not analytic, a typical case being the standard real mean square error objective function. The recent HR calculus is a step forward and provides a way to calculate derivatives and gradients of both analytic and non-analytic functions of quaternion variables; however, the HR calculus can become cumbersome in complex optimization problems due to the lack of rigorous product and chain rules, a consequence of the non-commutativity of quaternion algebra. To address this issue, we introduce the generalized HR (GHR) derivatives which employ quaternion rotations in a general orthogonal system and provide the left- and right-hand versions of the quaternion derivative of general functions. The GHR calculus also solves the long-standing problems of product and chain rules, mean-value theorem and Taylor's theorem in the quaternion field. At the core of the proposed GHR calculus is quaternion rotation, which makes it possible to extend the principle to other functional calculi in non-commutative settings. Examples in statistical learning theory and adaptive signal processing support the analysis.

1. INTRODUCTION

Quaternions have become a standard tool in many modern areas, including image processing [**1**,**2**], aerospace and satellite tracking [**3**], modelling of wind profile in renewable energy [**4**] and in the processing of polarized waves [**5**,**6**]. Compared to real vector algebra, quaternion algebra [**7**] has been shown to both reduce the number of parameters in the modelling and offer advantages in terms of functional simplicity and accuracy [**8**,**9**]. The most common optimization approach in applications is based on the gradient of the objective function; one such algorithm is the quaternion least mean square (QLMS) [**4**]. The objective functions in practical applications are typically based on the mean square error (MSE), a real function of quaternion variables, and are thus not analytic according to standard quaternion analysis [**10–12**]. This is a major obstacle to a more widespread use of quaternions in learning systems.

The existing ways to find the derivative of a real function $f(q)$ with respect to the unknown quaternion variable q are:

- *The pseudo-derivative,* which considers f as a function of the four real componentsq_a, q_b, q_c and q_d of the quaternion variable q, and then takes componentwise real derivatives with respect to the real variables q_a, q_b, q_c and q_d. In other words, the real-valued function f:H→R is treated as a real differentiable mapping between R4 and R. This leads to unnecessarily long expressions and is especially cumbersome and tedious in complex optimization problems and when dealing with nonlinear functions.

- *The HR calculus*, which is compact and elegant [**13**], as it finds formal derivatives of $f$with respect to the quaternion variables and their involutions in a direct way. This applies to both real functions of quaternion variables and nonlinear functions. This approach is based on the differentials of q,q^i,q^j,q^k, which are independent in the quaternion field, as shown in lemmas A.1 and A.2. The advantage of using HR derivatives is that the computations and analysis are kept in the quaternion domain rather than using quaternion-to-real transformations, and many algorithms can be readily extended from the complex to the quaternion domain.

Although the HR calculus is a significant step forward, the product and chain rules are not defined within the HR calculus, which complicates the calculation of derivatives of, for example, nonlinear quaternion functions. Other functional calculi [**10–12**,**14**] in quaternion analysis similarly suffer from this obstacle.

The aim of this work is to revisit the HR calculus [**13**] and to equip it with the product rule and chain rule in order to solve these long-standing problems in quaternion calculus. Motivated by the complex CR calculus [**15–17**], we first consider a general orthogonal system which, in conjunction with the HR calculus, introduces the generalized HR (GHR) calculus. The GHR calculus comprises both the left- and right-hand versions of quaternion derivative; these are necessary to consider due to non-commutativity of quaternion product. In particular, we show that for real functions of quaternion variables, such as the standard MSE objective function, the left and right GHR derivatives are identical. An important consequence of this property is that within the GHR calculus, the choice of the left/right GHR derivative is irrelevant for practical applications of quaternion optimization; this is currently a major source of confusion in the quaternion community. Another consequence of the novel product rule is that it not only enables the calculation of the GHR derivatives for general functions of quaternion variables, but also it is generic—if one function within the product is real-valued, this novel product rule degenerates into the traditional product rule, as shown in corollary 4.11. A family of chain rules is also introduced in order to calculate the derivatives of nonlinear functions of quaternion variables, which include complex- and real-valued functions as degenerate quaternion functions. Since at the core of the GHR calculus is the quaternion rotation, this approach can be naturally extended to other functional calculi in non-commutative settings. Finally, we revisit two fundamental theorems in quaternion calculus—the quaternion mean value theorem and quaternion Taylor's theorem—and derive them in a compact and generic form, based on the GHR derivatives. The GHR calculus therefore provides a solution to some long-standing mathematical problems [**18**] and promises a tool for a more widespread use of quaternions in practical applications. Illustrative examples in statistical signal processing support the analysis.

2. BACKGROUND ON QUATERNIONS

2.1 Quaternion algebra

Quaternions are an associative but not commutative algebra over R, defined as

$$\mathbb{H} = \text{span}\{1, i, j, k\} \triangleq \{q_a + iq_b + jq_c + kq_d \mid q_a, q_b, q_c, q_d \in \mathbb{R}\}, \tag{2.1}$$

where $\{1,i,j,k\}$ is a basis of H, and the imaginary units i,j and k satisfy $i^2=j^2=k^2=ijk=-1$, which implies $ij=k=-ji$, $jk=i=-kj$, $ki=j=-ik$. For any quaternion $q=qa+iqb+jqc+kqd=Sq+Vq$, the scalar (real) part is denoted by $qa=Sq=\mathrm{R}(q)$, whereas the vector part $Vq=\mathrm{I}(q)=iqb+jqc+kqd$ spans the three imaginary parts. For $p,q\in \mathrm{H}$, the quaternion product is given by $pq=SpSq-Vp\cdot Vq+SpVq+SqVp+Vp\times Vq$, where the symbols $'\cdot'$ and $'\times'$ denote, respectively, the standard inner product and vector product. The presence of vector product makes the quaternion product non-commutative, i.e. in general for $p,q\in \mathrm{H}$, $pq\neq qp$. The conjugate of a quaternion q is defined as $q^*=Sq-Vq$, while the conjugate of a quaternion product satisfies $(pq)^*=q^*p^*$. The modulus of a quaternion is defined as $|q|=\sqrt{qq*}$, and thus $|pq|=|p||q|$. The inverse of a quaternion $q\neq 0$ is $q^{-1}=q^*/|q|^2$ which yields an important consequence $(pq)^{-1}=q^{-1}p^{-1}$. If $|q|=1$, we call q a *unit*quaternion. A quaternion q is said to be *pure* if $\mathrm{R}(q)=0$. For pure quaternions, $q^*=-q$ and $q^2=-|q|^2$. Thus, a *pure unit* quaternion is a square root of -1; examples are the imaginary units i,$j$and k.

Quaternions can also be written in the *polar* form $q=|q|(\cos\theta+\hat{}q\sin\theta)$, where $\hat{}q=Vq/|Vq|$ is a pure unit quaternion and $\theta=\arccos(\mathrm{R}(q)/|q|)\in \mathrm{R}$ is the angle (or argument) of the quaternion. We shall next introduce the quaternion rotation and involution operators.

Definition 2.1 (quaternion rotation [19])

For any quaternion q, the transformation

$$q^{\mu} \triangleq \mu q \mu^{-1} \tag{2.2}$$

geometrically describes a three-dimensional rotation of the vector part of q by an angle 2θ about the vector part of μ, where $\mu=|\mu|(\cos\theta+\hat{}\mu\sin\theta)$ is any non-zero quaternion.

Properties of the quaternion rotation (see [**6**,**20**] used in this work) are

$$(pq)^{\mu} = p^{\mu} q^{\mu}, \quad pq = q^{p} p = q p^{(q^*)}, \quad \forall\, p,q \in \mathbb{H} \tag{2.3}$$

And

$$q^{\mu\nu} = (q^{\nu})^{\mu}, \quad q^{\mu *} \triangleq (q^*)^{\mu} = (q^{\mu})^* \triangleq q^{*\mu}, \quad \forall\, \nu, \mu \in \mathbb{H}. \qquad 2.4$$

Note that the representation in (2.1) can be generalized to a general orthogonal basis {1,*iμ*,*jμ*,*kμ*}, where the following properties hold [**19**]:

$$i^{\mu} i^{\mu} = j^{\mu} j^{\mu} = k^{\mu} k^{\mu} = i^{\mu} j^{\mu} k^{\mu} = -1. \qquad 2.5$$

Definition 2.2 (quaternion involution [21])

The involution of a quaternion *q* around a pure unit quaternion *η* is given by

$$q^{\eta} = \eta q \eta^{-1} = \eta q \eta^* = -\eta q \eta$$

and represents a rotation of *q* about *η* by *π*.

Of particular interest to this work are quaternion involutions around the imaginary units *i,j,k*, given by [**21**]

$$\left.\begin{aligned} q &= q_a + iq_b + jq_c + kq_d, \quad q^i = -iqi = q_a + iq_b - jq_c - kq_d \\ q^j &= -jqj = q_a - iq_b + jq_c - kq_d, \quad q^k = -kqk = q_a - iq_b - jq_c + kq_d, \end{aligned}\right\} \qquad 2.6$$

which allows us to express the four real-valued components of a quaternion *q* as [**13,21**]

$$q_a = \frac{1}{4}(q + q^i + q^j + q^k), \quad q_b = \frac{1}{4i}(q + q^i - q^j - q^k), \qquad 2.7$$

$$q_c = \frac{1}{4j}(q - q^i + q^j - q^k), \quad q_d = \frac{1}{4k}(q - q^i - q^j + q^k). \qquad 2.8$$

This is analogous to the complex case, where $x = \frac{1}{2}(z + z^*)$ and $y=-(i/2)(z-z^*)$ for any $z=x+iy \in \mathbb{C}$ [**22**]. Note that the quaternion conjugation operator $(\cdot)^*$ is also an involution and can be written in terms of q, q^i, q^j and q^k as

$$q^* = \frac{1}{2}(-q + q^i + q^j + q^k). \tag{2.9}$$

2.2 Analytic functions in H

To arrive at the notion of analytic (regular, monogenic) function in H, recall that due to the non-commutativity of quaternion product, there are two ways to express the quotient in the definition of quaternion derivative, as shown below.

Proposition 2.3 (Sudbery [10])

Let D⊆H be a simply connected domain of definition of the function f:D→H. If for any q∈D

$$\lim_{h\to 0}[(f(q+h) - f(q))h^{-1}] \tag{2.10}$$

exists in H, *then necessarily* $f(q)=\omega q+\lambda$ *for some* $\omega,\lambda\in$H. *If for any q∈D*

$$\lim_{h\to 0}[h^{-1}(f(q+h) - f(q))] \tag{2.11}$$

exists in H, *then necessarily* $f(q)=qv+\lambda$ *for some* $v,\lambda\in$H.

Proposition 2.3 is discussed in [**10,23**] and indicates that the traditional definitions of derivative in (2.10) and (2.11) are too restrictive and apply only to linear functions of quaternions. One attempt to relax this constraint is due to Fueter [**24,25**], whose analyticity condition is termed the Cauchy–Riemann–Fueter (CRF) equation, and is given by [**10,11**]

$$\left.\begin{aligned}&\text{Left CRF}: \quad \frac{\partial f}{\partial q_a}+\frac{\partial f}{\partial q_b}i+\frac{\partial f}{\partial q_c}j+\frac{\partial f}{\partial q_d}k=0\\&\text{Right CRF}: \quad \frac{\partial f}{\partial q_a}+i\frac{\partial f}{\partial q_b}+j\frac{\partial f}{\partial q_c}+k\frac{\partial f}{\partial q_d}=0.\end{aligned}\right\} \tag{2.12}$$

The limitations of the CRF condition were pointed out by Gentili & Struppa in [**14**,**26**], who showed that general polynomial functions (even the identity $f(q)=q$) satisfy neither the left CRF nor the right CRF. To further relax the analyticity condition, a *local* analyticity condition (LAC) was proposed in [**12**], by using the polar form of a quaternion, to give

$$\left.\begin{aligned}&\text{Left LAC}: \quad \frac{\partial f}{\partial q_a}+\left(q_b\frac{\partial f}{\partial q_a}+q_c\frac{\partial f}{\partial q_c}+q_d\frac{\partial f}{\partial q_d}\right)\frac{V_q}{|V_q|^2}=0\\&\text{Right LAC}: \quad \frac{\partial f}{\partial q_a}+\frac{V_q}{|V_q|^2}\left(q_b\frac{\partial f}{\partial q_a}+q_c\frac{\partial f}{\partial q_c}+q_d\frac{\partial f}{\partial q_d}\right)=0,\end{aligned}\right\} \tag{2.13}$$

where $q=qa+Vq$ and $Vq=iqb+jqc+kqd$. The theory of local analyticity is now well developed, and we refer the reader to [**14**,**26**,**27**] for the slice regular functions and to [**28**] for applications. More recent work in this area includes [**29**,**30**], and references therein. The advantage of the LAC is that both the polynomial functions of q and some elementary functions satisfy either the left LAC or the right LAC. However, in general, the products and compositions of two LAC functions f and$g$ no longer meet the local analytic condition. For example, if $f(q)=q$ and $g(q)=\omega q$, $\omega \in$H, then $f$and g satisfy the left LAC, but the product $fg=q\omega q$ does not satisfy the left LAC. The same applies for the right LAC, when the function g is written as $g(q)=q\omega$.

Quaternion derivative is defined only for analytic functions; however, in optimization it is often required that the objective function to be minimized or maximized is real-valued. A typical example is the mean square type objective function given by

$$J(\mathbf{w})=\|f(\mathbf{w})\|^2. \tag{2.14}$$

Note that according to the definition of analytic (regular) function given in [**10–12**,**14**,**26**,**27**], the function J is not analytic. In order to take its derivatives (but not limited to only such real quadratic functions), we can, however, use the HR calculus [**13**] which extends the complex CR calculus [**15**,**17**,**31**] to the quaternion field. This generalization is not trivial, and the rules of the CR and HR calculus are different; for more detail see §3.

Remark 2.4

It is important to note that the left (right) terminology introduced in (2.12), (2.13) and in the sequel differs from that in [**10–12**,**14**,**27**]. We adopt the use of the left (right) terminology based on the position of $\partial f/\partial q a$, $\partial f/\partial q b$, $\partial f/\partial q c$ and $\partial f/\partial q d$, rather than on the positions of imaginary unitsi,j,k. Although this is only a notational difference, we later show that the left derivatives (using this convenience) in definitions 3.2 and 4.1 are in this way equipped with a left constant rule (4.3), that is, the *left* constant comes out from the *left* derivative of product, and the left derivatives stand on the left side of the quaternion differential in (A 12) and (A 16). This also allows for a consistent, intuitive and physically meaningful use of terminology.

3. THE HR CALCULUS

Optimization problems involving quaternions arise in a number of applications in control theory, signal processing, robotics and biomechanics. Solutions often require a first- or second-order approximation of the objective function; however, real functions of quaternion variables are essentially non-analytic. The recently proposed HR calculus [**13**] solves these issues through the use of quaternion involutions. The HR derivatives (the derivation of HR calculus is given in appendix A) are introduced below.

Definition 3.1 (real-differentiability [10])

A function *f*(*q*)=*fa*(*qa*,*qb*,*qc*,*qd*)+ *ifb*(*qa*,*qb*,*qc*,*qd*)+*jfc*(*qa*,*qb*,*qc*,*qd*)+*kfd* (*qa*,*qb*,*qc*,*qd*) is called real differentiable when *fa*(*qa*,*qb*,*qc*,*qd*), *fb*(*qa*,*qb*,*qc*,*qd*), *fc*(*qa*,*qb*,*qc*,*qd*) and *fd*(*qa*,*qb*,*qc*,*qd*) are differentiable as functions of real variables *qa*,*qb*,*qc* and *qd*.

Definition 3.2 (the HR derivatives [13])

If *f*:H→H is real-differentiable, then the formal left HR derivatives of the function *f* with respect to

{*q*,*qi*,*qj*,*qk*} and {*q**,*qi**,*qj**,*qk**} are defined as

$$\begin{bmatrix} \frac{\partial f}{\partial q}, & \frac{\partial f}{\partial q^*} \\ \frac{\partial f}{\partial q^i}, & \frac{\partial f}{\partial q^{i*}} \\ \frac{\partial f}{\partial q^j}, & \frac{\partial f}{\partial q^{j*}} \\ \frac{\partial f}{\partial q^k}, & \frac{\partial f}{\partial q^{k*}} \end{bmatrix}^{\mathrm{T}} = \frac{1}{4} \begin{bmatrix} \frac{\partial f}{\partial q_a}, & \frac{\partial f}{\partial q_a} \\ \frac{\partial f}{\partial q_b}, & -\frac{\partial f}{\partial q_b} \\ \frac{\partial f}{\partial q_c}, & -\frac{\partial f}{\partial q_c} \\ \frac{\partial f}{\partial q_d}, & -\frac{\partial f}{\partial q_d} \end{bmatrix}^{\mathrm{T}} \begin{bmatrix} 1 & 1 & 1 & 1 \\ -i & -i & i & i \\ -j & j & -j & j \\ -k & k & k & -k \end{bmatrix} \qquad 3.1$$

and the formal right HR derivatives of the function *f* are defined as

$$\begin{bmatrix} \frac{\partial_r f}{\partial q}, & \frac{\partial_r f}{\partial q^*} \\ \frac{\partial_r f}{\partial q^i}, & \frac{\partial_r f}{\partial q^{i*}} \\ \frac{\partial_r f}{\partial q^j}, & \frac{\partial_r f}{\partial q^{j*}} \\ \frac{\partial_r f}{\partial q^k}, & \frac{\partial_r f}{\partial q^{k*}} \end{bmatrix} = \frac{1}{4} \begin{bmatrix} 1 & -i & -j & -k \\ 1 & -i & j & k \\ 1 & i & -j & k \\ 1 & i & j & -k \end{bmatrix} \begin{bmatrix} \frac{\partial f}{\partial q_a}, & \frac{\partial f}{\partial q_a} \\ \frac{\partial f}{\partial q_b}, & -\frac{\partial f}{\partial q_b} \\ \frac{\partial f}{\partial q_c}, & -\frac{\partial f}{\partial q_c} \\ \frac{\partial f}{\partial q_d}, & -\frac{\partial f}{\partial q_d} \end{bmatrix}, \qquad 3.2$$

where $\partial f/\partial qa$, $\partial f/\partial qb$, $\partial f/\partial qc$ and $\partial f/\partial qd$ are the partial derivatives of *f* with respect to *qa*, *qb*, *qc* andqd.

Remark 3.3

It is important to note that the right HR derivatives exist if and only if the left HR derivatives also exist. The only difference between the left HR derivatives and the right HR derivatives is in the position of the partial derivative $\partial f/\partial qa$, $\partial f/\partial qb$, $\partial f/\partial qc$ and $\partial f/\partial qd$. Within the left HR derivatives, $\partial f/\partial qa$,$\partial f/\partial qb$,$\partial f/\partial qc$ and $\partial f/\partial qd$ stand on the left side and imaginary units *i,j,k* on the right side (cf. (3.1)). It is exactly the opposite case for right HR derivatives. Note that the terms $\partial f/\partial qa$,$\partial f/\partial qb$,$\partial f/\partial qc$ and $\partial f/\partial qd$ cannot swap positions with the imaginary units *i,j,k* because of the non-commutative nature of quaternion product.

3.1 The validity of the traditional product rule

A straightforward use of the HR derivatives may become too tedious for complicated functions, for example, for the power function $f(q)=qn$. This is because the HR calculus does not satisfy the traditional product rule which would simplify the calculation. Indeed, for two quaternion functions$f(q)$ and $g(q)$, in general we have $\partial(fg)/\partial q \neq f(\partial g/\partial q)+\partial f/\partial qg$. We shall illustrate this difficulty through the following two examples.

Example 3.4

Find the HR derivative of the function f:H→H given by

$$f(q) = q^2 = q_a^2 - (q_b^2 + q_c^2 + q_d^2) + 2q_a(iq_b + jq_c + kq_d), \qquad 3.3$$

where $q=qa+iqb+jqc+kqd$, $qa,qb,qc,qd \in$R.

Solution: By definition 3.2, the left HR derivative of q^2 becomes

$$\frac{\partial(q^2)}{\partial q} = \frac{1}{4}\left(\frac{\partial q^2}{\partial q_a} - \frac{\partial q^2}{\partial q_b}i - \frac{\partial q^2}{\partial q_c}j - \frac{\partial q^2}{\partial q_d}k\right)$$
$$= \frac{1}{4}(2q_a + 2(iq_b + jq_c + kq_d) - (-2q_b + 2q_a i)i - (-2q_c + 2q_a j)j - (-2q_d + 2q_a k)k)$$
$$= \frac{1}{4}(8q_a + 4q_b i + 4q_c j + 4q_d k) = q + \Re(q).$$

3.4

Alternatively, $q(\partial q/\partial q)+(\partial q/\partial q)q=2q$. This shows the traditional product rule is not valid.

Example 3.5

Find the HR derivative of the function

$$f(q) = |q|^2 = qq^*.$$ 3.5

Solution: By definition 3.2, the HR derivative of $|q|^2$ is

$$\frac{\partial(|q|^2)}{\partial q} = \frac{1}{4}\left(\frac{\partial|q|^2}{\partial q_a} - \frac{\partial|q|^2}{\partial q_b}i - \frac{\partial|q|^2}{\partial q_c}j - \frac{\partial|q|^2}{\partial q_d}k\right) = \frac{1}{4}(2q_a - 2q_b i - 2q_c j - 2q_d k) = \frac{1}{2}q^*,$$

3.6

while $q(\partial q^*/\partial q)+(\partial q/\partial q)q^*=-q/2+q^*$, and thus the traditional product rule is not valid.

Remark 3.6

Examples 3.4 and 3.5 show that the left HR derivatives do not admit the traditional product rule. Similarly, the traditional product rule is not applicable for the right HR derivative in definition 3.2.

4. THE GENERALIZATION OF HR CALCULUS: GENERALIZED HR DERIVATIVES

We now introduce the novel GHR derivatives which comprise both the product and chain rules. This is archived by replacing the basis $\{1,i,j,k\}$ in definition 3.2 with a general orthogonal basis $\{1,i\mu,j\mu,k\mu\}$, see also (2.5). The derivation of the GHR calculus is similar to that of the HR calculus in appendix A and is omitted for space considerations.

Definition 4.1 (the generalized HR derivatives)

If f:H→H is real-differentiable, then the left GHR derivatives of the function f with respect to $q\mu$ and $q\mu^*$ ($\mu\neq 0,\mu\in$H) are defined as

$$\left.\begin{aligned}\frac{\partial f}{\partial q^{\mu}} &= \frac{1}{4}\left(\frac{\partial f}{\partial q_a} - \frac{\partial f}{\partial q_b} i^{\mu} - \frac{\partial f}{\partial q_c} j^{\mu} - \frac{\partial f}{\partial q_d} k^{\mu}\right) \\ \frac{\partial f}{\partial q^{\mu *}} &= \frac{1}{4}\left(\frac{\partial f}{\partial q_a} + \frac{\partial f}{\partial q_b} i^{\mu} + \frac{\partial f}{\partial q_c} j^{\mu} + \frac{\partial f}{\partial q_d} k^{\mu}\right),\end{aligned}\right\} \quad 4.1$$

while the right GHR derivatives are defined as

$$\left.\begin{aligned}\frac{\partial_r f}{\partial q^{\mu}} &= \frac{1}{4}\left(\frac{\partial f}{\partial q_a} - i^{\mu}\frac{\partial f}{\partial q_b} - j^{\mu}\frac{\partial f}{\partial q_c} - k^{\mu}\frac{\partial f}{\partial q_d}\right) \\ \frac{\partial_r f}{\partial q^{\mu *}} &= \frac{1}{4}\left(\frac{\partial f}{\partial q_a} + i^{\mu}\frac{\partial f}{\partial q_b} + j^{\mu}\frac{\partial f}{\partial q_c} + k^{\mu}\frac{\partial f}{\partial q_d}\right),\end{aligned}\right\} \quad 4.2$$

where $q=qa+iqb+jqc+kqd$, $qa,qb,qc,qd\in$R, $\partial f/\partial qa$, $\partial f/\partial qb$, $\partial f/\partial qc$ and $\partial f/\partial qd$ are the partial derivatives of f with respect to qa, qb, qc and qd, while the set $\{1,i\mu,j\mu,k\mu\}$ is an orthogonal basis of H.

Remark 4.2

The GHR derivatives are more concise and physically more intuitive than the HR derivatives, which are a special case of the GHR derivatives for $\mu=\{1,i,j,k\}$ (definitions 3.2 and 4.1). The concept of the

GHR derivative can also be applied to other orthogonal systems, such as $\{1,\eta,\eta',\eta''\}$ in **[32]**.

Proposition 4.3 (constant rule)

Let f:H→H *be real-differentiable. Then, the following*

holds:

$$\frac{\partial(\nu f)}{\partial q^{\mu}} = \nu\frac{\partial f}{\partial q^{\mu}}, \quad \frac{\partial(f\nu)}{\partial q^{\mu}} = \frac{\partial f}{\partial q^{\nu\mu}}\nu \qquad 4.3$$

And

$$\frac{\partial_r(f\nu)}{\partial q^{\mu}} = \frac{\partial_r f}{\partial q^{\mu}}\nu, \quad \frac{\partial_r(\nu f)}{\partial q^{\mu}} = \nu\frac{\partial_r f}{\partial q^{\nu^*\mu}}, \qquad 4.4$$

where $\mu,\nu\in$H *are non-zero quaternion constants.*

Proof.

By the definition of the left GHR derivative in (4.1), we have

$$\begin{aligned}\frac{\partial(\nu f)}{\partial q^{\mu}} &= \frac{1}{4}\left(\frac{\partial(\nu f)}{\partial q_a} - \frac{\partial(\nu f)}{\partial q_b}i^{\mu} - \frac{\partial(\nu f)}{\partial q_c}j^{\mu} - \frac{\partial(\nu f)}{\partial q_d}k^{\mu}\right)\\ &= \frac{1}{4}\left(\nu\frac{\partial f}{\partial q_a} - \nu\frac{\partial f}{\partial q_b}i^{\mu} - \nu\frac{\partial f}{\partial q_c}j^{\mu} - \nu\frac{\partial f}{\partial q_d}k^{\mu}\right) = \nu\frac{\partial f}{\partial q^{\mu}}.\end{aligned} \qquad 4.5$$

Using the equality $(q^{\mu})^{\nu} = q^{\nu\mu}$ in (2.4), the second equality of (4.3) is proved as

$$\frac{\partial(f\nu)}{\partial q^{\mu}}=\frac{1}{4}\left(\frac{\partial(\nu f)}{\partial q_a}-\frac{\partial(\nu f)}{\partial q_b}i^{\mu}-\frac{\partial(\nu f)}{\partial q_c}j^{\mu}-\frac{\partial(\nu f)}{\partial q_d}k^{\mu}\right)$$
$$=\frac{1}{4}\left(\frac{\partial f}{\partial q_a}\nu-\frac{\partial f}{\partial q_b}\nu i^{\mu}-\frac{\partial f}{\partial q_c}\nu j^{\mu}-\nu\frac{\partial f}{\partial q_d}\nu k^{\mu}\right)$$
$$=\frac{1}{4}\left(\frac{\partial f}{\partial q_a}-\frac{\partial f}{\partial q_b}i^{\nu\mu}-\frac{\partial f}{\partial q_c}j^{\nu\mu}-\nu\frac{\partial f}{\partial q_d}k^{\nu\mu}\right)\nu=\frac{\partial f}{\partial q^{\nu\mu}}\nu. \qquad 4.6$$

Hence, (4.3) immediately follows, and (4.4) can be proved in a similar way. ▪

Remark 4.4

It is important to note that if a function f is premultiplied by a constant η in the first equality of (4.3), then the left GHR derivative of the product is equal to the left GHR derivative of $f$premultiplied by the constant, but not for postmultiplication. In other words, the left constant νcan come out from the derivative of the product; for this reason we refer to (4.1) as the left GHR derivative. The equalities in (4.4) complement those in (4.3). Thus, we refer to the derivatives in (4.2) as the right GHR derivatives, denoted by ∂r in order to distinguish them from the left GHR derivatives in (4.1).

Proposition 4.5 (conjugate rule)

Let f:H→H be real-differentiable. Then, the following holds:

$$\frac{\partial f}{\partial q^{\mu}}=\left(\frac{\partial_r f^*}{\partial q^{\mu *}}\right)^* \quad and \quad \frac{\partial f}{\partial q^{\mu *}}=\left(\frac{\partial_r f^*}{\partial q^{\mu}}\right)^*. \qquad 4.7$$

Proof.

By the definition of the right GHR derivative in (4.2), we have

$$\left(\frac{\partial_r f^*}{\partial q^{\mu *}}\right)^*=\frac{1}{4}\left(\frac{\partial f^*}{\partial q_a}+i^{\mu}\frac{\partial f^*}{\partial q_b}+j^{\mu}\frac{\partial f^*}{\partial q_c}+k^{\mu}\frac{\partial f^*}{\partial q_d}\right)^*. \qquad 4.8$$

Using the equalities $(pq)^*=q^*p^*$ and $(q^*)\mu=(q\mu)^*$ in (2.4), the above equality becomes

$$\left(\frac{\partial_r f^*}{\partial q^{\mu *}}\right)^* = \frac{1}{4}\left(\frac{\partial f}{\partial q_a} - \frac{\partial f}{\partial q_b} i^{\mu} - \frac{\partial f}{\partial q_c} j^{\mu} - \frac{\partial f}{\partial q_d} k^{\mu}\right) = \frac{\partial f}{\partial q^{\mu}}. \qquad 4.9$$

Hence, the first part of (4.7) follows, and the second part can be proved in a similar way. ▪

Proposition 4.6 (rotation rule)

Let f:H→H *be real-differentiable. Then the following holds*:

$$\left(\frac{\partial f}{\partial q^{\mu}}\right)^{\nu} = \frac{\partial f^{\nu}}{\partial q^{\nu\mu}} \quad and \quad \left(\frac{\partial f}{\partial q^{\mu *}}\right)^{\nu} = \frac{\partial f^{\nu}}{\partial q^{\nu\mu *}}. \qquad .4.10$$

Proof. **Using the equalities in (2.4) and the left GHR derivative in (4.1), we have**

$$\begin{aligned}\left(\frac{\partial f}{\partial q^{\mu}}\right)^{\nu} &= \frac{1}{4}\left(\frac{\partial f}{\partial q_a} - \frac{\partial f}{\partial q_b} i^{\mu} - \frac{\partial f}{\partial q_c} j^{\mu} - \frac{\partial f}{\partial q_d} k^{\mu}\right)^{\nu} \\ &= \frac{1}{4}\left(\frac{\partial f^{\nu}}{\partial q_a} - \frac{\partial f^{\nu}}{\partial q_b} i^{\nu\mu} - \frac{\partial f^{\nu}}{\partial q_c} j^{\nu\mu} - \frac{\partial f^{\nu}}{\partial q_d} k^{\nu\mu}\right) = \frac{\partial f^{\nu}}{\partial q^{\nu\mu}}.\end{aligned} \qquad 4.11$$

Hence, the first part of (4.10) follows, and the second part can be proved in a similar way. ▪

Corollary 4.7

Let f:H→H *be real-differentiable. Then, the following holds*:

$$\left(\frac{\partial f}{\partial q^{\eta}}\right)^{\eta} = \frac{\partial f^{\eta}}{\partial q} \quad and \quad \left(\frac{\partial f}{\partial q^{\eta *}}\right)^{\eta} = \frac{\partial f^{\eta}}{\partial q^*}, \quad \forall\, \eta \in \{1, i, j, k\}. \qquad 4.12$$

Proof. **Since $\partial f/\partial q\eta\eta=\partial f/\partial q$ for any η?{1,*i*,*j*,*k*}, then the proof follows directly from proposition 4.6.** ▪

Proposition 4.8

Let f:H→R *be real-differentiable. Then, the following holds*:

$$\frac{\partial f}{\partial q^{\mu}}=\left(\frac{\partial f}{\partial q^{\mu *}}\right)^{*}=\frac{\partial_r f}{\partial q^{\mu}} \quad and \quad \frac{\partial f}{\partial q^{\mu *}}=\left(\frac{\partial f}{\partial q^{\mu}}\right)^{*}=\frac{\partial_r f}{\partial q^{\mu *}}. \qquad 4.13$$

Proof. **Since *f* is real-valued, its partial derivatives $\partial f/\partial qa$, $\partial f/\partial qb$, $\partial f/\partial qc$ and $\partial f/\partial qd$ are real numbers, which yields $(\partial f/\partial \xi)^*=\partial f/\partial \xi$, where $\xi\{qa,qb,qc,qd\}$. Using the equality $(q\mu)^*=(q^*)\mu$ in (2.4) and the left GHR derivative in (4.1), we have**

$$\begin{aligned}\left(\frac{\partial f}{\partial q^{\mu *}}\right)^{*} &= \frac{1}{4}\left(\frac{\partial f}{\partial q_a}+\frac{\partial f}{\partial q_b}i^{\mu}+\frac{\partial f}{\partial q_c}j^{\mu}+\frac{\partial f}{\partial q_d}k^{\mu}\right)^{*}\\ &= \frac{1}{4}\left(\frac{\partial f}{\partial q_a}-i^{\mu}\frac{\partial f}{\partial q_b}-j^{\mu}\frac{\partial f}{\partial q_c}-k^{\mu}\frac{\partial f}{\partial q_d}\right)=\frac{\partial_r f}{\partial q^{\mu}}\\ &= \frac{1}{4}\left(\frac{\partial f}{\partial q_a}-\frac{\partial f}{\partial q_b}i^{\mu}-\frac{\partial f}{\partial q_c}j^{\mu}-\frac{\partial f}{\partial q_d}k^{\mu}\right)=\frac{\partial f}{\partial q^{\mu}}.\end{aligned} \qquad 4.14$$

Hence, the first part of (4.13) follows, and the second part can be proved in a similar way. ▪

Remark 4.9

From the identity (4.13), observe that the left GHR derivative is equal to the right GHR derivative if the function *f* is real-valued. This result is instrumental for practical applications of quaternion optimization, where the objective function (or cost function) is frequently real-valued. By virtue of the GHR calculus, the choice of the left/right GHR derivative therefore becomes obsolete as shown in (4.13). In the sequel, without loss in generality we shall mainly focus on the left GHR derivatives.

4.1 The novel product rule

We now introduce a novel product rule into quaternion analysis and show that the traditional product rule is a special case of the proposed product rule in corollary 4.11.

Theorem 4.10 (product rule of left GHR)

If the functions f,g:H→H are real-differentiable, then so too is their product fg, so that

$$\frac{\partial(fg)}{\partial q^{\mu}}=f\frac{\partial g}{\partial q^{\mu}}+\frac{\partial f}{\partial q^{g\mu}}g \quad and \quad \frac{\partial(fg)}{\partial q^{\mu*}}=f\frac{\partial g}{\partial q^{\mu*}}+\frac{\partial f}{\partial q^{g\mu*}}g, \tag{4.15}$$

where $\partial f/\partial q^{g\mu}$ and $\partial f/\partial q^{g\mu*}$ can be obtained by replacing μ with $g\mu$ in definition 4.1.

Proof. **The proof is given in appendix B.** ▪

Corollary 4.11 ***If the functions f:H→H and g:H→R are real-differentiable, then their product fg satisfies the traditional product rule∂***

$$\frac{\partial(fg)}{\partial q^{\mu}}=f\frac{\partial g}{\partial q^{\mu}}+\frac{\partial f}{\partial q^{\mu}}g \quad and \quad \frac{\partial(fg)}{\partial q^{\mu*}}=f\frac{\partial g}{\partial q^{\mu*}}+\frac{\partial f}{\partial q^{\mu*}}g, \tag{4.16}$$

where $\partial f/\partial q^{\mu}$ and $\partial f/\partial q^{\mu*}$ are the left GHR derivatives in definition 4.1.

Proof. **Since $q^{g\mu}=q^{\mu}$ and $q^{g\mu*}=q^{\mu*}$ for a real function g:H→R, the corollary follows.** ▪

We now present some GHR derivatives of nonlinear quaternion functions enabled by the GHR calculus. These are very useful in applications, such as in nonlinear adaptive filters and quaternion-valued neural networks.

Example 4.12 (split-quaternion function)

Find the GHR derivative of a split-quaternion function *f*:H→H given by

$$f(q) = \varphi(q_a) + j\varphi(q_b) + j\varphi(q_c) + k\varphi(q_d), \quad 4.17$$

where *q*=*qa*+*iqb*+*jqc*+*kqd*∈R and *φ*:R→R is a real-valued differentiable function.

Solution: By the definition of the left GHR derivatives in (4.1), it follows that

$$\begin{aligned}\frac{\partial f(q)}{\partial q} &= \frac{1}{4}\left(\frac{\partial f(q)}{\partial q_a} - \frac{\partial f(q)}{\partial q_a}i - \frac{\partial f(q)}{\partial q_a}j - \frac{\partial f(q)}{\partial q_a}k\right)\\ &= \frac{1}{4}(\varphi'(q_a) + \varphi'(q_b) + \varphi'(q_c) + \varphi'(q_d))\end{aligned} \quad 4.18$$

And

$$\begin{aligned}\frac{\partial f(q)}{\partial q^*} &= \frac{1}{4}\left(\frac{\partial f(q)}{\partial q_a} + \frac{\partial f(q)}{\partial q_a}i + \frac{\partial f(q)}{\partial q_a}j + \frac{\partial f(q)}{\partial q_a}k\right)\\ &= \frac{1}{4}(\varphi'(q_a) - \varphi'(q_b) - \varphi'(q_c) - \varphi'(q_d)).\end{aligned} \quad 4.19$$

This shows that the GHR derivatives of the split-quaternion function are real-valued.

Example 4.13 (power function)

Find the GHR derivative of the power function *f*:H→H given by

$$f(q) = q^n, \quad 4.20$$

where *n* is any positive integer.

Solution: Using the product rule in theorem 4.10, it follows that

$$\frac{\partial q^n}{\partial q^\mu}\mu = \frac{\partial (qq^{n-1})}{\partial q^\mu}\mu = q\frac{\partial q^{n-1}}{\partial q^\mu}\mu + \frac{\partial q}{\partial q^{q^{n-1}\mu}}q^{n-1}\mu = q\frac{\partial q^{n-1}}{\partial q^\mu}\mu + \Re(q^{n-1}\mu), \quad 4.21$$

where the term $(\partial q/\partial q\mu)\mu$, given in **table 1**, was used in the last equality. Note that the above expression is recurrent in $(\partial qn/\partial q\mu)\mu$. Upon expanding this expression and using the initial condition $(\partial q/\partial q\mu)\mu=\mathrm{R}(\mu)$, this yields

$$\frac{\partial q^n}{\partial q^\mu}\mu = \sum_{m=1}^{n} q^{n-m}\Re(q^{m-1}\mu). \quad 4.22$$

In a similar manner, we have

$$\frac{\partial q^n}{\partial q^{\mu*}}\mu = \frac{\partial (qq^{n-1})}{\partial q^{\mu*}}\mu = q\frac{\partial q^{n-1}}{\partial q^{\mu*}}\mu + \frac{\partial q}{\partial q^{q^{n-1}\mu*}}q^{n-1}\mu = q\frac{\partial q^{n-1}}{\partial q^{\mu*}}\mu - \frac{1}{2}(q^{n-1}\mu)^*, \quad 4.23$$

which is equivalent to

$$\frac{\partial q^n}{\partial q^{\mu*}}\mu = -\frac{1}{2}\sum_{m=1}^{n} q^{n-m}(q^{m-1}\mu)^*. \quad 4.24$$

Example 4.14 (exponential function)(power function)

Find the GHR derivative of the function *f*:H→H given by

$$\exp(q) \triangleq \sum_{n=0}^{+\infty}\frac{q^n}{n!}. \quad 4.25$$

Solution: From (4.22), it follows that

$$\frac{\partial \exp(q)}{\partial q^\mu}\mu = \sum_{n=0}^{+\infty}\frac{1}{n!}\sum_{m=1}^{n} q^{n-m}\Re(q^{m-1}\mu). \quad 4.26$$

Table 1. Summary of the rules for the left GHR derivatives.

$f(q)$	$\frac{\partial f}{\partial q^{\mu}}\mu$	$\frac{\partial f}{\partial q^{\mu *}}\mu$	note
q	$\Re(\mu)$	$-\frac{1}{2}\mu^*$	—
ωq	$\omega\Re(\mu)$	$-\frac{1}{2}\omega\mu^*$	$\forall\,\omega \in \mathbb{H}$
$q\nu$	$\Re(\nu\mu)$	$-\frac{1}{2}(\nu\mu)^*$	$\forall\,\nu \in \mathbb{H}$
$\omega q\nu + \lambda$	$\omega\Re(\nu\mu)$	$-\frac{1}{2}\omega(\nu\mu)^*$	$\forall\,\omega,\nu,\lambda \in \mathbb{H}$
q^*	$-\frac{1}{2}\mu^*$	$\Re(\mu)$	—
ωq^*	$-\frac{1}{2}\omega\mu^*$	$\omega\Re(\mu)$	$\forall\,\omega \in \mathbb{H}$
$q^*\nu$	$-\frac{1}{2}(\nu\mu)^*$	$\Re(\nu\mu)$	$\forall\,\nu \in \mathbb{H}$
$\omega q^*\nu + \lambda$	$-\frac{1}{2}\omega(\nu\mu)^*$	$\omega\Re(\nu\mu)$	$\forall\,\omega,\nu,\lambda \in \mathbb{H}$
q^{-1}	$-q^{-1}\Re(q^{-1}\mu)$	$\frac{1}{2}q^{-1}\mu^*(q^*)^{-1}$	—
$(q^*)^{-1}$	$\frac{1}{2}(q^*)^{-1}\mu^*q^{-1}$	$-(q^*)^{-1}\Re((q^*)^{-1}\mu)$	—
$(\omega q\nu + \lambda)^{-1}$	$-f\omega\Re(\nu f\mu)$	$\frac{1}{2}f\omega(\nu f\mu)^*$	$\forall\,\omega,\nu,\lambda \in \mathbb{H}$
q^2	$q\Re(\mu) + \Re(q\mu)$	$-\frac{1}{2}q\mu^* - \frac{1}{2}(q\mu)^*$	—
$(q^*)^2$	$-\frac{1}{2}q^*\mu^* - \frac{1}{2}(q^*\mu)^*$	$q^*\Re(\mu) + \Re(q^*\mu)$	—
$(\omega q\nu + \lambda)^2$	$g\omega\Re(\nu\mu) + \omega\Re(\nu g\mu)$	$-\frac{1}{2}g\omega(\nu\mu)^* - \frac{1}{2}\omega(\nu g\mu)^*$	$g = \omega q\nu + \lambda$
$(\omega q^*\nu + \lambda)^2$	$-\frac{1}{2}g\omega(\nu\mu)^* - \frac{1}{2}\omega(\nu g\mu)^*$	$g\omega\Re(\nu\mu) + \omega\Re(\nu g\mu)$	$g = \omega q^*\nu + \lambda$
$\Re(q)$	$\frac{1}{4}\mu$	$\frac{1}{4}\mu$	—
$\Re(\omega q\nu + \lambda)$	$\frac{1}{4}\mu\nu\omega$	$\frac{1}{4}\mu\omega^*\nu^*$	$\forall\,\omega,\nu,\lambda \in \mathbb{H}$
$\Re(\omega q^*\nu + \lambda)$	$\frac{1}{4}\mu\omega^*\nu^*$	$\frac{1}{4}\mu\nu\omega$	$\forall\,\omega,\nu,\lambda \in \mathbb{H}$
$\|q\|$	$\frac{1}{4\|q\|}\mu q^*$	$\frac{1}{4\|q\|}\mu q$	—
$\|q\|^2$	$\frac{1}{2}\mu q^*$	$\frac{1}{2}\mu q$	—
$\|\omega q\nu + \lambda\|$	$\frac{1}{4\|g\|}\mu\nu g^*\omega$	$\frac{1}{4\|g\|}\mu\omega^* g\nu^*$	$g = \omega q\nu + \lambda$
$\|\omega q^*\nu + \lambda\|$	$\frac{1}{4\|g\|}\mu\omega^* g\nu^*$	$\frac{1}{4\|g\|}\mu\nu g^*\omega$	$g = \omega q^*\nu + \lambda$
$\|\omega q\nu + \lambda\|^2$	$\frac{1}{2}\mu\nu g^*\omega$	$\frac{1}{2}\mu\omega^* g\nu^*$	$g = \omega q\nu + \lambda$
$\|\omega q\nu + \lambda\|^2$	$\frac{1}{2}\mu\nu g^*\omega$	$\frac{1}{2}\mu\omega^* g\nu^*$	$g = \omega q\nu + \lambda$
$\|\omega q^*\nu + \lambda\|^2$	$\frac{1}{2}\mu\omega^* g\nu^*$	$\frac{1}{2}\mu\nu g^*\omega$	$g = \omega q^*\nu + \lambda$

Table 1. *(Continued.)*

$f(q)$	$\frac{\partial f}{\partial q^{\mu}}\mu$	$\frac{\partial f}{\partial q^{\mu*}}\mu$	note
$\frac{q}{\lvert q\rvert}$	$\frac{1}{\lvert q\rvert}\Re(\mu)-\frac{1}{4\lvert q\rvert^3}q\mu q^*$	$-\frac{1}{2\lvert q\rvert}\mu^*-\frac{1}{4\lvert q\rvert^3}q\mu q$	—
$\frac{q^*}{\lvert q\rvert}$	$-\frac{1}{2\lvert q\rvert}\mu^*-\frac{1}{4\lvert q\rvert^3}q^*\mu q^*$	$\frac{1}{\lvert q\rvert}\Re(\mu)-\frac{1}{4\lvert q\rvert^3}q^*\mu q$	—
$\frac{\omega q\nu+\lambda}{\lvert\omega q\nu+\lambda\rvert}$	$\frac{\omega}{2\lvert g\rvert}\Re(\nu\mu)+\frac{g}{4\lvert g\rvert^3}\nu^*(\omega^* g\mu)^*$	$-\frac{\omega}{4\lvert g\rvert}(\nu\mu)^*-\frac{g}{2\lvert g\rvert^3}\nu^*\Re(\omega^* g\mu)$	$g=\omega q\nu+\lambda$
$\frac{\omega q^*\nu+\lambda}{\lvert\omega q^*\nu+\lambda\rvert}$	$-\frac{\omega}{2\lvert g\rvert}(\nu\mu)^*-\frac{f}{\lvert g\rvert}\frac{\partial\lvert g\rvert}{\partial q^{\mu}}\mu$	$\frac{\omega}{\lvert g\rvert}\Re(\nu\mu)-\frac{f}{\lvert g\rvert}\frac{\partial\lvert g\rvert}{\partial q^{\mu*}}\mu$	$g=\omega q^*\nu+\lambda$

In a similar manner, we have

$$\frac{\partial\exp(q)}{\partial q^{\mu*}}\mu=-\frac{1}{2}\sum_{n=0}^{+\infty}\frac{1}{n!}\sum_{m=1}^{n}q^{n-m}(q^{m-1}\mu)^*. \tag{4.27}$$

Remark 4.15

The exponential function is the most important elementary function, as both trigonometric functions and hyperbolic functions can be expressed in terms of the exponential function. The elementary function in example 4.14 is a power series, and it does not change the direction of the vector part of quaternion. Therefore, such elementary functions can swap positions with a quaternion q, i.e. $f(q)q=qf(q)$, giving an important property, $f^*(q)=f(q^*)$, which can be used in practical applications, such as quaternion neural networks [**33**] and quaternion nonlinear adaptive filters [**28**]. It is important to note that if a quaternion variable q degenerates into a real variable x in the definitions of elementary functions in this subsection, then the GHR derivatives simplify into the standard real derivatives, e.g. the GHR derivative of the power function in (4.22) will become nx^{n-1}. Therefore, the GHR derivatives are a generalized form of the real derivatives and the real derivatives are a special case of the GHR derivatives.

4.2 The chain rule

Another advantage of the GHR derivatives is that they admit the chain rule, which is formulated in the following theorem.

Theorem 4.16 (chain rule of left GHR)

Let S⊆H and let g:S→H be real-differentiable at an interior point q of the set S. Let T⊆H be such that g(q)∈T for all q∈S. Assume that f:T→H is real-differentiable at an interior point g(q)∈T, then the composite function f(g(q)) satisfies the following chain rules:

$$\left.\begin{aligned}\frac{\partial f(g(q))}{\partial q^{\mu}}&=\frac{\partial f}{\partial g^{\nu}}\frac{\partial g^{\nu}}{\partial q^{\mu}}+\frac{\partial f}{\partial g^{\nu i}}\frac{\partial g^{\nu i}}{\partial q^{\mu}}+\frac{\partial f}{\partial g^{\nu j}}\frac{\partial g^{\nu j}}{\partial q^{\mu}}+\frac{\partial f}{\partial g^{\nu k}}\frac{\partial g^{\nu k}}{\partial q^{\mu}}\\\frac{\partial f(g(q))}{\partial q^{\mu *}}&=\frac{\partial f}{\partial g^{\nu}}\frac{\partial g^{\nu}}{\partial q^{\mu *}}+\frac{\partial f}{\partial g^{\nu i}}\frac{\partial g^{\nu i}}{\partial q^{\mu *}}+\frac{\partial f}{\partial g^{\nu j}}\frac{\partial g^{\nu j}}{\partial q^{\mu *}}+\frac{\partial f}{\partial g^{\nu k}}\frac{\partial g^{\nu k}}{\partial q^{\mu *}}\end{aligned}\right\} \quad 4.28$$

And

$$\left.\begin{aligned}\frac{\partial f(g(q))}{\partial q^{\mu}}&=\frac{\partial f}{\partial g^{\nu *}}\frac{\partial g^{\nu *}}{\partial q^{\mu}}+\frac{\partial f}{\partial g^{\nu i*}}\frac{\partial g^{\nu i*}}{\partial q^{\mu}}+\frac{\partial f}{\partial g^{\nu j*}}\frac{\partial g^{\nu j*}}{\partial q^{\mu}}+\frac{\partial f}{\partial g^{\nu k*}}\frac{\partial g^{\nu k*}}{\partial q^{\mu}}\\\frac{\partial f(g(q))}{\partial q^{\mu *}}&=\frac{\partial f}{\partial g^{\nu *}}\frac{\partial g^{\nu *}}{\partial q^{\mu *}}+\frac{\partial f}{\partial g^{\nu i*}}\frac{\partial g^{\nu i*}}{\partial q^{\mu *}}+\frac{\partial f}{\partial g^{\nu j*}}\frac{\partial g^{\nu j*}}{\partial q^{\mu *}}+\frac{\partial f}{\partial g^{\nu k*}}\frac{\partial g^{\nu k*}}{\partial q^{\mu *}},\end{aligned}\right\} \quad 4.29$$

where $\mu,\nu\in$ H, $\mu\nu\neq 0$, and $\partial f/\partial g\nu=\partial f(g)/\partial g\nu$ and $\partial f/\partial g\nu^{*}=\partial f(g)/\partial g\nu^{*}$ are the left GHR derivatives.

Proof. **The proof of theorem 4.16 is given in appendix C.** ▪

Theorem 4.16 is also valid for complex-valued and real-valued composite functions of quaternion variables, as stated in the following two corollaries, the proofs of which are similar to that of theorem 4.16, and are thus omitted.

Corollary 4.17

Let S⊆H and let g:S→C be real-differentiable at an interior point q of the set S. Let T⊆C be such that g(q)∈T for all q∈S. Assume that f:T→C is real-differentiable at an interior point g(q)∈T, then the left GHR derivatives of the composite function f(g(q)) are as follows:

$$\frac{\partial f(g(q))}{\partial q^{\mu}} = \frac{\partial f}{\partial g}\frac{\partial g}{\partial q^{\mu}} + \frac{\partial f}{\partial g^*}\frac{\partial g^*}{\partial q^{\mu}} \quad and \quad \frac{\partial f(g(q))}{\partial q^{\mu *}} = \frac{\partial f}{\partial g}\frac{\partial g}{\partial q^{\mu *}} + \frac{\partial f}{\partial g^*}\frac{\partial g^*}{\partial q^{\mu *}},$$

4.30

where μ∈H,μ≠0, and ∂f/∂g=∂f(g)/∂g and ∂f/∂g=∂f(g)/∂g* are the complex CR derivatives within the CR calculus.*

Corollary 4.18 ***Let S⊆H and let g:S→R be real-differentiable at an interior point q of the set S. Let T⊆R be such that g(q)∈T for all q∈S. Assume that f:T→R is real-differentiable at an interior point f(q)∈T, then the left GHR derivatives of the composite function f(g(q)) are as follows:***

$$\frac{\partial f(g(q))}{\partial q^{\mu}} = f'(g)\frac{\partial g}{\partial q^{\mu}} \quad and \quad \frac{\partial f(g(q))}{\partial q^{\mu *}} = f'(g)\frac{\partial g}{\partial q^{\mu *}},$$

4.31

where μ∈H,μ≠0 and f′(g) is the real derivative of a real-valued function.

Theorem 4.19 (chain rule of right GHR)

Let S⊆H and let g:S→H be real-differentiable at an interior point q of the set S. Let T⊆H be such that g(q)∈T for all q∈S. Assume that f:T→H is real-differentiable at an inner point g(q)∈T, then the right GHR derivatives of the composite function f(g(q)) are as follows:

$$\left.\begin{aligned}\frac{\partial_r f(g(q))}{\partial q^{\mu}} &= \frac{\partial_r g^{\nu}}{\partial q^{\mu}}\frac{\partial_r f}{\partial g^{\nu}} + \frac{\partial_r g^{\nu i}}{\partial q^{\mu}}\frac{\partial_r f}{\partial g^{\nu i}} + \frac{\partial_r g^{\nu j}}{\partial q^{\mu}}\frac{\partial_r f}{\partial g^{\nu j}} + \frac{\partial_r g^{\nu k}}{\partial q^{\mu}}\frac{\partial_r f}{\partial g^{\nu k}} \\ \frac{\partial_r f(g(q))}{\partial q^{\mu *}} &= \frac{\partial_r g^{\nu}}{\partial q^{\mu *}}\frac{\partial_r f}{\partial g^{\nu}} + \frac{\partial_r g^{\nu i}}{\partial q^{\mu *}}\frac{\partial_r f}{\partial g^{\nu i}} + \frac{\partial_r g^{\nu j}}{\partial q^{\mu *}}\frac{\partial_r f}{\partial g^{\nu j}} + \frac{\partial_r g^{\nu k}}{\partial q^{\mu *}}\frac{\partial_r f}{\partial g^{\nu k}}\end{aligned}\right\} \qquad 4.32$$

And

$$\left.\begin{aligned}\frac{\partial_r f(g(q))}{\partial q^{\mu}} &= \frac{\partial_r g^{\nu *}}{\partial q^{\mu}}\frac{\partial_r f}{\partial g^{\nu *}} + \frac{\partial_r g^{\nu i *}}{\partial q^{\mu}}\frac{\partial_r f}{\partial g^{\nu i *}} + \frac{\partial_r g^{\nu j *}}{\partial q^{\mu}}\frac{\partial_r f}{\partial g^{\nu j *}} + \frac{\partial_r g^{\nu k *}}{\partial q^{\mu}}\frac{\partial_r f}{\partial g^{\nu k *}} \\ \frac{\partial_r f(g(q))}{\partial q^{\mu *}} &= \frac{\partial_r g^{\nu *}}{\partial q^{\mu *}}\frac{\partial_r f}{\partial g^{\nu *}} + \frac{\partial_r g^{\nu i *}}{\partial q^{\mu *}}\frac{\partial_r f}{\partial g^{\nu i *}} + \frac{\partial_r g^{\nu j *}}{\partial q^{\mu *}}\frac{\partial_r f}{\partial g^{\nu j *}} + \frac{\partial_r g^{\nu k *}}{\partial q^{\mu *}}\frac{\partial_r f}{\partial g^{\nu k *}},\end{aligned}\right\} \qquad 4.33$$

where $\mu,\nu \in \mathbb{H}, \mu\nu \neq 0$*, and* $\partial_r f/\partial g^{\nu} = \partial_r f(g)/\partial g^{\nu}$ *and* $\partial_r f/\partial g^{\nu *} = \partial_r f(g)/\partial g^{\nu *}$ *are the right GHR derivatives.*

Proof. **The proof of theorem 4.19 is similar to that of theorem 4.16 and is thus omitted.** ▪

4.3 Mean value theorem

The mean value theorem is one of the most important tools in calculus, and we next introduce its compact version for general quaternion-valued functions of quaternion variables.

Theorem 4.20 (mean value theorem of left kind)

Consider a continuous function $f: S \subseteq \mathbb{H} \to \mathbb{H}$ *for which the left GHR derivatives exist and are continuous in the set S. Then, for any* $q_0, q_1 \in S$ *for which the segment joining them also lies in S, we have*

$$f(q_1) - f(q_0) = \sum_{\mu} \int_0^1 \frac{\partial f(q_0 + t\lambda)}{\partial q^{\mu}} \lambda^{\mu}\, dt = \sum_{\mu} \int_0^1 \frac{\partial f(q_0 + t\lambda)}{\partial q^{\mu *}} \lambda^{\mu *}\, dt, \qquad 4.34$$

where $\mu \in \{1,i,j,k\}$*,* $\lambda = q_1 - q_0$*, and* $\partial f(q_0+t\lambda)/\partial q^{\mu} = \partial f(q)/\partial q^{\mu}|_{q=q_0+t\lambda}$ *is the left GHR derivative as in definition 4.1.*

Proof.

Denote $F(t)=f(g(t))$, where $g(t)=q_0+t\lambda$ and $0\leq t\leq 1$. Then $F(t)$ is continuous on [0,1] and has derivatives in (0,1). By theorem 4.16, the derivative of $F(t)$ is

$$\begin{aligned}F'(t) &= \frac{\partial f(g)}{\partial g}\lambda + \frac{\partial f(g)}{\partial g^i}\lambda^i + \frac{\partial f(g)}{\partial g^j}\lambda^j + \frac{\partial f(g)}{\partial g^k}\lambda^k \\ &= \frac{\partial f(q)}{\partial q}\bigg|_{q=q_0+t\lambda}\lambda + \frac{\partial f(q)}{\partial q^i}\bigg|_{q=q_0+t\lambda}\lambda^i + \frac{\partial f(q)}{\partial q^j}\bigg|_{q=q_0+t\lambda}\lambda^j + \frac{\partial f(q)}{\partial q^k}\bigg|_{q=q_0+t\lambda}\lambda^k \\ &= \frac{\partial f(q_0+t\lambda)}{\partial q}\lambda + \frac{\partial f(q_0+t\lambda)}{\partial q^i}\lambda^i + \frac{\partial f(q_0+t\lambda)}{\partial q^j}\lambda^j + \frac{\partial f(q_0+t\lambda)}{\partial q^k}\lambda^k.\end{aligned}$$

.4.35

Upon substituting (4.35) into $F(1)-F(0)=\int 10F'(t)\,dt$, with $F(0)=f(q_0)$ and $F(1)=f(q_1)$, the theorem follows. The second equality can be proved in a similar manner. ▪

Corollary 4.21 *Consider a continuous function f:S⊆H→R for which the left GHR derivatives exist and are continuous in the set S. Then, for any $q_0,q_1\in S$ for which the segment joining them also lies in S, we have*

$$f(q_1)-f(q_0) = 4\int_0^1 \Re\left(\frac{\partial f(q_0+t\lambda)}{\partial q}\lambda\right)dt = 4\int_0^1 \Re\left(\frac{\partial f(q_0+t\lambda)}{\partial q^*}\lambda^*\right)dt, \quad 4.36$$

where $\lambda=q_1-q_0$ and $\partial f(q_0+t\lambda)/\partial q\mu=\partial f(q)/\partial q\mu|_{q=q_0+t\lambda}$ is the left GHR derivative as in definition 4.1.

Proof.

Function f is real-valued, and therefore $\partial f/\partial q=\partial f\eta/\partial q$, where $\eta\in\{1,i,j,k\}$. From (2.3) and (4.10), we now have

$$\frac{\partial f}{\partial q^\eta}\lambda^\eta = \frac{\partial f^\eta}{\partial q^\eta}\lambda^\eta = \left(\frac{\partial f}{\partial q}\lambda\right)^\eta. \quad 4.37$$

The corollary then follows from (2.7) and theorem 4.20, while the second equality can be derived using R(pq)=R($p*q*$). ▪

For λ which is sufficiently small in the modulus, the right-hand side of (4.34) can be approximated as

$$f(q_1) - f(q_0) \approx \sum_{\mu \in \{1,i,j,k\}} \frac{\partial f(q_0)}{\partial q^\mu} \lambda^\mu. \tag{4.38}$$

If the left GHR derivatives of f are Lipschitz continuous in the vicinity of q and q_1, with the Lipschitz constant L, we can estimate the error in this approximation as

$$\left| f(q_1) - f(q_0) - \sum_{\mu \in \{1,i,j,k\}} \frac{\partial f(q_0)}{\partial q^\mu} \lambda^\mu \right| \leq \sum_{\mu \in \{1,i,j,k\}} \left| \int_0^1 \left(\frac{\partial f(q_0 + t\lambda)}{\partial q^\mu} - \frac{\partial f(q_0)}{\partial q^\mu} \right) \lambda^\mu \, \mathrm{d}t \right| \leq 4 \int_0^1 Lt|\lambda|^2 \, \mathrm{d}t = 2L|\lambda|^2. \tag{4.39}$$

4.4 Taylor's theorem

We can now introduce a novel, rigorous version of Taylor's theorem for quaternion-valued functions of quaternion variables, as a generalization of the standard univariate Taylor's theorem.

Lemma 4.22 (Apostol [34]) *Consider a (k+1)-times continuously differentiable function f:D⊆R→R. If x∈D, then*

$$f(x_0 + h) = f(x_0) + f'(x_0)h + f''(x_0)\frac{h^2}{2} + \cdots + f^{(k)}(x_0)\frac{h^k}{k!} + R_k, \tag{4.40}$$

where the remainder Rk is given by

$$R_k = \int_{x_0}^{x_0+h} \frac{f^{(k+1)}(t)}{k!}(x_0 + h - t)^k \, dt. \tag{4.41}$$

Theorem 4.23 (Taylor's theorem of left kind)

Consider a third-order continuous real-differentiable function f:S⊆H→H. If $q_0, q_0+\lambda \in S$ such that the segment joining them also lies in S, then

$$\begin{aligned} f(q_0 + \lambda) - f(q_0) &= \sum_{\mu} \frac{\partial f(q_0)}{\partial q^{\mu}} \lambda^{\mu} + \frac{1}{2} \sum_{\mu,\nu} \frac{\partial^2 f(q_0)}{\partial q^{\nu} \partial q^{\mu}} \lambda^{\nu} \lambda^{\mu} + O(\lambda^3) \\ &= \sum_{\mu} \frac{\partial f(q_0)}{\partial q^{\mu *}} \lambda^{\mu *} + \frac{1}{2} \sum_{\mu,\nu} \frac{\partial^2 f(q_0)}{\partial q^{\nu *} \partial q^{\mu *}} \lambda^{\nu *} \lambda^{\mu *} + O(\lambda^3) \quad \text{as } \lambda \to 0, \end{aligned} \tag{4.42}$$

where $\mu,\nu \in \{1,i,j,k\}$, and $\partial^2 f/\partial q\nu \partial q\mu$ and $\partial^2 f/\partial q\nu^* \partial q\mu^*$ are the second-order left GHR derivatives.

Proof. Define an auxiliary function $g(t)=f(q_0+t\lambda)$ with $0 \le t \le 1$. Using the chain rule in theorem 4.16, we obtain

$$\left. \begin{aligned} g'(t) &= \sum_{\mu} \frac{\partial f(q_0 + t\lambda)}{\partial q^{\mu}} \lambda^{\mu} \\ g''(t) &= \sum_{\mu,\nu} \frac{\partial^2 f(q_0 + t\lambda)}{\partial q^{\nu} \partial q^{\mu}} \lambda^{\nu} \lambda^{\mu}, \quad g'''(t) = \sum_{\mu,\nu,\eta} \frac{\partial^3 f(q_0 + t\lambda)}{\partial q^{\eta} \partial q^{\nu} \partial q^{\mu}} \lambda^{\eta} \lambda^{\nu} \lambda^{\mu}, \end{aligned} \right\} \tag{4.43}$$

where $\mu,\nu,\eta \in \{1,i,j,k\}$. The second-order Taylor polynomial in lemma 4.22 then gives

$$g(1) = g(0) + g'(0) + \tfrac{1}{2} g''(0) + R_2 \tag{4.44}$$

which is equivalent to

$$f(q_0+\lambda)=f(q_0)+\sum_{\mu}\frac{\partial f(q_0)}{\partial q^{\mu}}\lambda^{\mu}+\frac{1}{2}\sum_{\mu,\nu}\frac{\partial^2 f(q_0)}{\partial q^{\nu}\partial q^{\mu}}\lambda^{\nu}\lambda^{\mu}+R_2, \tag{4.45}$$

Where

$$R_2=\int_0^1\frac{(1-t)^2}{2}g'''(t)\,\mathrm{d}t=\int_0^1\frac{(1-t)^2}{2}\sum_{\mu,\nu,\eta}\frac{\partial^3 f(q_0+t\lambda)}{\partial q^{\eta}\partial q^{\nu}\partial q^{\mu}}\lambda^{\eta}\lambda^{\nu}\lambda^{\mu}\,\mathrm{d}t. \tag{4.46}$$

This integral contains three factors of λ, while the remaining factors are bounded. Therefore, $R_2$is of the order $|\lambda|^3$, making the fraction $|R_2|/|\lambda|^3$ bounded as $\lambda\to 0$. Hence, the first equality of the theorem follows, and the second equality can be proved in the same way. ▪

Corollary 4.24

Consider a third-order continuous real-differentiable function $f:S\subseteq\mathbb{H}\to\mathbb{R}$. If $q_0,q_0+\lambda\in S$ such that the segment joining them also lies in S, then

$$\begin{aligned}f(q_0+\lambda)-f(q_0)&=4\Re\left(\frac{\partial f(q_0)}{\partial q}\lambda\right)+2\sum_{\nu}\Re\left(\frac{\partial^2 f(q_0)}{\partial q^{\nu}\partial q}\lambda^{\nu}\lambda\right)+O(|\lambda|^3)\\&=4\Re\left(\frac{\partial f(q_0)}{\partial q^*}\lambda^*\right)+2\sum_{\nu}\Re\left(\frac{\partial^2 f(q_0)}{\partial q^{\nu*}\partial q^*}\lambda^{\nu*}\lambda^*\right)+O(|\lambda|^3)\quad \text{as } \lambda\to 0,\end{aligned} \tag{4.47}$$

where $\nu\in\{1,i,j,k\}$, and $\partial^2 f/\partial q^{\nu}\partial q$ and $\partial^2 f/\partial q^{\nu*}\partial q^*$ are the second-order left GHR derivatives.

Proof. Using the result for $\partial R(q)/\partial q$ in table 1 and the chain rule in theorem 4.16, this corollary is proved similarly to the proof of corollary 4.21. ▪

Theorem 4.25 (Taylor's theorem of centre kind)

Consider a third-order continuous real-differentiable function $f:S\subseteq H\rightarrow R$. If $q_0, q_0+\lambda \in S$ such that the segment joining them also lies in S, then

$$\begin{aligned} f(q_0+\lambda)-f(q_0) &= \sum_{\mu} \frac{\partial f(q_0)}{\partial q^{\mu}} \lambda^{\mu} + \frac{1}{2} \sum_{\mu,\nu} \lambda^{\mu *} \frac{\partial^2 f(q_0)}{\partial q^{\nu} \partial q^{\mu *}} \lambda^{\nu} + O(\lambda^3) \\ &= \sum_{\mu} \frac{\partial f(q_0)}{\partial q^{\mu *}} \lambda^{\mu *} + \frac{1}{2} \sum_{\mu,\nu} \lambda^{\mu} \frac{\partial^2 f(q_0)}{\partial q^{\nu *} \partial q^{\mu}} \lambda^{\nu *} + O(\lambda^3) \quad \text{as } \lambda \rightarrow 0, \end{aligned} \tag{4.48}$$

where $\mu,\nu \in \{1,i,j,k\}$, and $\partial^2 f/\partial q\nu \partial q\mu^*$ and $\partial^2 f/\partial q\nu \partial q\mu^*$ are the second-order left GHR derivatives.

Proof. Define an auxiliary function $g(t)=f(q_0+t\lambda)$ with $0 \leq t \leq 1$. Using the chain rule in theorem 4.19 and using the property in (4.13), we obtain

$$g'(t) = \sum_{\mu} \lambda^{\mu *} \frac{\partial_r f(q_0+t\lambda)}{\partial q^{\mu *}} = \sum_{\mu} \lambda^{\mu *} \frac{\partial f(q_0+t\lambda)}{\partial q^{\mu *}}. \tag{4.49}$$

Using the constant rule (4.3), we arrive at

$$g''(t) = \sum_{\mu} \lambda^{\mu *} \left(\frac{\partial f(q_0+t\lambda)}{\partial q^{\mu *}} \right)' = \sum_{\mu,\nu} \lambda^{\mu *} \frac{\partial^2 f(q_0+t\lambda)}{\partial q^{\nu} \partial q^{\mu *}} \lambda^{\nu}, \tag{4.50}$$

where $\mu,\nu \in \{1,i,j,k\}$. The rest of the proof is almost the same as that in theorem 4.23 and is thus omitted. ▪

Remark 4.26

The Taylor expansion in theorem 4.23 is concisely expressed using the GHR derivatives. This is different from the Taylor expansion given by Schwartz [**18**], which decomposes a quaternion $q$into two mutually perpendicular quaternions in a local coordinate system. In contrast, our approach treats the quaternion q as an augmented quaternion based on quaternion involutions [**21**]. Schwartz has also stated that his Taylor expansion would cause trouble when the function has terms $q\omega q$, where ω is a general quaternion. Note that there are no such restrictions in theorem 4.23, which only requires the real-differentiability condition in definition 3.1, that is, the functions $f(q)$ should admit the partial derivatives with respect to the four real componentsq_a,q_b,q_c and q_d.

5. APPLICATIONS OF THE GENERALIZED HR CALCULUS

We now illustrate the utility of the GHR calculus in optimization, statistics, signal processing, machine learning and other application areas.

5.1 Derivation of the widely linear quaternion least mean square algorithm

The widely linear QLMS (WL-QLMS) algorithm is based on the quaternion widely linear model$y(n)=\mathbf{w}^T(n)\mathbf{p}(n)$ which deals with the generality of quaternion signals (both proper and improper) [**8**,**22**,**35**], where $\mathbf{p}=(\mathbf{x}^T(n),\mathbf{x}^{iT}(n),\mathbf{x}^{jT}(n),\mathbf{x}^{kT}(n))^T$ is the augmented input vector and $\mathbf{w}=(\mathbf{h}^T(n),\mathbf{g}^T(n),\mathbf{u}^T(n),\mathbf{v}^T(n))^T$ is the associated weight (parameter) vector. The cost function to be minimized is a real-valued function of quaternion variables

$$J(n) = |e(n)|^2, \qquad 5.1$$

where $e(n)=d(n)-y(n)$ is the error between the desired signal $d(n)$ and the filter output $y(n)$. The weight update of WL-QLMS is then given by

$$\Delta \mathbf{w}(n) = \mathbf{w}(n+1) - \mathbf{w}(n) = -\alpha \nabla_{\mathbf{w}^*} J(n) = -\alpha \left(\frac{\partial J(n)}{\partial \mathbf{w}^*} \right)^{\mathrm{T}}, \tag{5.2}$$

where α is the step size, $(\cdot)^{\mathrm{T}}$ denotes the transpose and the gradient $\nabla_{\mathbf{w}*} J(n)$ defines the direction of the maximum rate of change of J [**13,36**]. By the product rule within the GHR calculus, given in theorem 4.10, we have

$$\frac{\partial J}{\partial \mathbf{w}^*} = e^* \frac{\partial e}{\partial \mathbf{w}^*} + \frac{\partial e^*}{\partial \mathbf{w}^{e*}} e, \tag{5.3}$$

where the time index 'n' is omitted. Note the convention that $\partial f/\partial \mathbf{w}$ is a row vector whose nth element is $\partial f/\partial w_n$. The above GHR derivatives are calculated as

$$\left.\begin{aligned} \frac{\partial e}{\partial \mathbf{w}^*} &= \frac{\partial(d - \mathbf{w}^{\mathrm{T}}\mathbf{p})}{\partial \mathbf{w}^*} = -\frac{\partial(\mathbf{w}^{\mathrm{T}}\mathbf{p})}{\partial \mathbf{w}^*} = \frac{1}{2}\mathbf{p}^H \\ \frac{\partial e^*}{\partial \mathbf{w}^{e*}} e &= \frac{\partial(d^* - \mathbf{p}^H\mathbf{w}^*)}{\partial \mathbf{w}^{e*}} e = -\frac{\partial(\mathbf{p}^H\mathbf{w}^*)}{\partial \mathbf{w}^{e*}} e = -\mathbf{p}^H \Re(e), \end{aligned}\right\} \tag{5.4}$$

where the terms $\partial(qv)/\partial q^*$ and $\partial(\omega q^*)/\partial q^{\mu *}$ are given in **table 1** and are used in the last equalities in the expressions above. Substituting (5.4) into (5.3) yields

$$\frac{\partial J}{\partial \mathbf{w}^*} = \frac{1}{2} e^* \mathbf{p}^H - \mathbf{p}^H \Re(e) = \left(\frac{1}{2} e^* - \Re(e) \right) \mathbf{p}^H = -\frac{1}{2} e \mathbf{p}^H. \tag{5.5}$$

Finally, the update of the adaptive weight vector of WL-QLMS becomes

$$\mathbf{w}(n+1) = \mathbf{w}(n) + \alpha e(n) \mathbf{p}^*(n), \tag{5.6}$$

where the constant 12 in (5.5) is absorbed into α.

Remark 5.1

There are many variations of WL-QLMS algorithms, such as the WL-QLMS algorithms based on variants $\{\mathbf{x}^*, \mathbf{x}i^*, \mathbf{x}j^*, \mathbf{x}k^*\}$, $\{\mathbf{x}\mu, \mathbf{x}\mu i, \mathbf{x}\mu j, \mathbf{x}\mu k\}$ and $\{\mathbf{x}^*, \mathbf{x}\mu i^*, \mathbf{x}\mu j^*, \mathbf{x}\mu k^*\}$. Note that if we start from$y(n)=\mathbf{w}H(n)\mathbf{p}(n)$, the final update rule would become $\mathbf{w}(n+1)=\mathbf{w}(n)+\alpha\mathbf{p}(n)e^*(n)$.

5.2 Derivation of quaternion nonlinear adaptive filtering algorithms

Tools of the GHR calculus allow us to concisely derive quaternion nonlinear adaptive filtering algorithms, a basis for fast-growing area of quaternion learning system. The same real-valued quadratic cost function as in real LMS and complex LMS is used, that is

$$J(n) = |e(n)|^2 = e_a^2(n) + e_b^2(n) + e_c^2(n) + e_d^2(n), \tag{5.7}$$

where $e(n)=ea(n)+ieb(n)+jec(n)+ked(n)$, $e(n)=d(n)-\Phi(s(n))$, $s(n)=\mathbf{w}^{\mathrm{T}}(n)\mathbf{x}(n)$, and Φ is the quaternion nonlinearity. The weight update is given by

$$\Delta\mathbf{w}(n) = \mathbf{w}(n+1) - \mathbf{w}(n) = -\alpha\nabla_{\mathbf{w}^*}J(n) = -\alpha\left(\frac{\partial J(n)}{\partial \mathbf{w}^*}\right)^{\mathrm{T}}, \tag{5.8}$$

where $\alpha>0$ is the real step size, $(\cdot)^{\mathrm{T}}$ denotes the transpose, and the gradient $\nabla w{*}J(n)$ defines the direction of the maximum rate of change of J [**13,36**]. By using the chain rule in theorem4.16, the above gradient can be calculated as

$$\frac{\partial J}{\partial \mathbf{w}^*} = \sum_{\mu\in\{1,i,j,k\}} \frac{\partial J}{\partial s^{\mu*}} \frac{\partial s^{\mu*}}{\partial \mathbf{w}^*}, \tag{5.9}$$

where time index 'n' is omitted for convenience. Note the convention that $\partial f/\partial \mathbf{w}$ is a row vector whose nth element is $\partial f/\partial wn$. Using the term $\partial(\omega q^*)/\partial q^*$ in **table 1**, we have

$$\frac{\partial s^*}{\partial \mathbf{w}^*} = \frac{\partial(\mathbf{x}^H \mathbf{w}^*)}{\partial \mathbf{w}^*} = \mathbf{x}^H. \qquad 5.10$$

Upon applying the second equality in (4.12) and using the term $\partial(\omega q^*)/\partial q\mu^*\mu$ in **table 1**, this yields

$$\frac{\partial s^{\mu *}}{\partial \mathbf{w}^*} = \left(\frac{\partial s^*}{\partial \mathbf{w}^{\mu *}}\right)^{\mu} = -\mu \frac{\partial(\mathbf{x}^H \mathbf{w}^*)}{\partial \mathbf{w}^{\mu *}} \mu = -\mu \mathbf{x}^H \Re(\mu) = 0, \quad \forall \mu \in \{i, j, k\}. \qquad 5.11$$

Using the chain rule in corollary 4.18, we have

$$\begin{aligned} \frac{\partial J}{\partial s^*} &= 2e_a \frac{\partial e_a}{\partial s^*} + 2e_b \frac{\partial e_b}{\partial s^*} + 2e_c \frac{\partial e_c}{\partial s^*} + 2e_d \frac{\partial e_d}{\partial s^*} \\ &= -2e_a \frac{\partial \Phi_a(s)}{\partial s^*} - 2e_b \frac{\partial \Phi_b(s)}{\partial s^*} - 2e_c \frac{\partial \Phi_c(s)}{\partial s^*} - 2e_d \frac{\partial \Phi_d(s)}{\partial s^*}. \end{aligned} \qquad 5.12$$

Upon substituting (5.10)–(5.12) into (5.9), we arrive at

$$\frac{\partial J}{\partial \mathbf{w}^*} = -2\left(e_a \frac{\partial \Phi_a(s)}{\partial s^*} + e_b \frac{\partial \Phi_b(s)}{\partial s^*} + e_c \frac{\partial \Phi_c(s)}{\partial s^*} + e_d \frac{\partial \Phi_d(s)}{\partial s^*}\right) \mathbf{x}^H. \qquad 5.13$$

Finally, the weight update for this quaternion nonlinear adaptive filtering algorithm becomes

$$\Delta \mathbf{w}(n) = \alpha \left(e_a(n) \frac{\partial \Phi_a(s(n))}{\partial s^*(n)} + e_b(n) \frac{\partial \Phi_b(s(n))}{\partial s^*(n)} + e_c(n) \frac{\partial \Phi_c(s(n))}{\partial s^*(n)} + e_d(n) \frac{\partial \Phi_d(s(n))}{\partial s^*(n)}\right) \mathbf{x}^*(n), \qquad 5.14$$

where the constant 2 in (5.13) is absorbed into the step size α.

For illustration, consider an example where Φ is a nonlinear split-quaternion function$\Phi(s)=\varphi(sa)+i\varphi(sb)+j\varphi(sc)+\varphi(sd)$ and φ:R→R is a real-valued differentiable function. Then,

$$\frac{\partial \Phi_a(s)}{\partial s^*} = \frac{\partial \varphi(s_a)}{\partial s^*} = \frac{1}{4}\varphi'(s_a). \qquad 5.15$$

In a similar manner, we have

$$\frac{\partial \Phi_b(s)}{\partial s^*} = \frac{i}{4}\varphi'(s_b), \quad \frac{\partial \Phi_c(s)}{\partial s^*} = \frac{j}{4}\varphi'(s_c) \quad \text{and} \quad \frac{\partial \Phi_d(s)}{\partial s^*} = \frac{k}{4}\varphi'(s_d). \qquad 5.16$$

The weight update for such a quaternion nonlinear adaptive filtering algorithm becomes

$$\Delta \mathbf{w}(n) = \alpha(e_a(n)\varphi'(s_a(n)) + ie_b(n)\varphi'(s_b(n)) + je_c(n)\varphi'(s_c(n)) + ke_d(n)\varphi'(s_d(n)))\mathbf{x}^*(n), \qquad 5.17$$

where the constant 14 in (5.15) and (5.16) is absorbed into the step size α.

Another example is when Φ is a quaternion linear function $\Phi(s)=s$, that is $\varphi(x)=x$ in (5.17). Then, the update of the adaptive weight vector within the QLMS algorithm becomes

$$\mathbf{w}(n+1) = \mathbf{w}(n) + \alpha e(n)\mathbf{x}^*(n) \qquad 5.18$$

illustrating the generic nature of the GHR calculus.

Remark 5.2

From (5.17) and (5.18), we note that quaternion linear and nonlinear adaptive filtering algorithms have been developed in a unified form. This also shows that the GHR calculus gives us much more freedom in the design, as the nonlinear function Φ is not required to satisfy the odd-symmetry condition [**28,37**]. We can derive many other algorithms, such as the augmented quaternion nonlinear gradient descent algorithm [**38**], in the same way. In the interest of space, we leave this to the interested reader.

Remark 5.3

The QLMS algorithm (5.18) is different from the QLMS in [**4**,**13**], due to the use of different product rule. The traditional product rule was used in [**4**,**13**] to simplify the derivation; however, our examples in §3a illustrate that the traditional product rule is not applicable to the HR derivatives. On the other hand, the chain rules within the GHR calculus result in the QLMS in (5.18), which has the same generic form as that of the complex LMS [**39**]. For the performance comparison and steady-state analysis of the existing QLMS algorithms, we refer the reader to [**40**] for more details.

6. CONCLUSION

A novel and rigorous framework for the efficient computation of quaternion derivatives, referred to as the GHR calculus, has been established. The GHR methodology has been shown to greatly relax the existence conditions for the derivatives of general nonlinear functions of quaternion variables, and to simplify the calculation of quaternion derivatives through its novel product and chain rules. We have shown that, unlike the existing quaternion derivatives, the GHR calculus is general and can be used for both analytic and non-analytic functions of quaternion variables. The core of the GHR calculus is the use of quaternion rotations in order to overcome the non-commutativity of quaternion product, and the use of quaternion involutions to obtain an elegant quaternion basis. Through the analysis and examples, the proposed framework has been shown to allow for real- and complex-valued optimization algorithms to be extended to the quaternion field in a generic, compact and intuitive way. Application case studies in statistical signal processing and learning systems demonstrate the effectiveness of the proposed GHR framework.

AUTHORS' CONTRIBUTIONS

D.X. designed the study, carried out the derivations and analysis and drafted the manuscript; C.J. helped with the extensions of HR calculus and the analysis; C.C.T. helped with the application case studies and proofread the manuscript; D.P.M. conceived and coordinated the study, was involved in the design and analysis, and helped in drafting the manuscript. All authors gave final approval for publication.

COMPETING INTERESTS

The authors have no competing interests.

FUNDING

D.X. was supported by the National Natural Science Foundation of China (no. 61301202), and by the Research Fund for the Doctoral Program of Higher Education of China (no. 20122304120028). D.P.M. was supported by EPSRC grant no. EP/H026266/1.

ACKNOWLEDGEMENTS

The authors wish to thank the anonymous reviewers for their insightful comments and suggestions on how to improve this work.

REFERENCES

1. Pei SC, Cheng CM. 1999 Color image processing by using binary quaternion-moment-preserving thresholding technique. IEEE Trans. Image Process. **8**, 614–628.

2. Bulow T, Sommer G. 2001 Hypercomplex signals: a novel extension of the analytic signal to the multidimensional case. IEEE Trans. Signal Process. **49**, 2844–2852.

3. Fortuna L, Muscato G, Xibilia M. 2001 A comparison between HMLP and HRBF for attitude control. IEEE Trans. Neural Netw. **12**, 318–328.

4. Took CC, Mandic DP. 2009 The quaternion LMS algorithm for adaptive filtering of hypercomplex processes.IEEE Trans. Signal Process. **57**, 1316–1327.

5. Le Bihan N, Mars J. 2004 Singular value decomposition of quaternion matrices: a new tool for vector-sensor signal processing. Signal Process. **84**, 1177–1199.

6. Buchholz S,

7. Le Bihan N. 2008 Polarized signal classification by complex and quaternionic multi-layer perceptrons. Int. J. Neural Syst. **18**, 75–85.

8. Hamilton WR. 1847 On quaternions. Proc. R. Irish Acad. **4**, 1–16.

9. Xia Y, Jahanchahi C, Mandic DP. 2015 Quaternion-valued echo state networks. IEEE Trans. Neural Netw. Learn. Syst. **26**, 663–673.

10. Mizoguchi T, Yamada I. 2014 An algebraic translation of Cayley-Dickson linear systems and its applications to online learning. IEEE Trans. Signal Process. **62**, 1438–1453.

11. Sudbery A. 1979 Quaternionic analysis. Math. Proc. Camb. Phil. Soc. **85**, 199–225.

12. Deavours CA. 1973 The quaternion calculus. Amer. Math. Mon. **80**, 995–1008. De Leo S, Rotelli PP. 2003 Quaternionic analyticity. App. Math. Lett. **16**, 1077–1081.

13. Mandic DP, Jahanchahi C, Took CC. 2011 A quaternion gradient operator and its applications. IEEE Signal Process. Lett. **18**, 47–50. ⍰

14. Gentili G, Struppa DC. 2006 A new approach to Cullen-regular functions of a quaternionic variable. C. R. Math. Acad. Sci. Paris **342**, 741–744. ⍰

15. Brandwood D. 1983 A complex gradient operator and its application in adaptive array theory. IEEE Commun. Radar Signal Process. **130**, 11–16.

16. Mandic DP, Goh SL. 2009 *Complex valued nonlinear adaptive filters: noncircularity, widely linear and neural models*. London, UK: John Wiley and Sons Ltd.

17. Kreutz-Delgado K. 2009 The complex gradient operator and the CR Calculus. (http://arxiv.org/abs/0906.4835)

18. Schwartz C. 2009 Calculus with a quaternionic variable. J. Math. Phys. **50**, 013523.

19. Ward JP. 1997 *Quaternions and Cayley numbers: algebra and applications*. London, UK: Kluwer Academic Publishers.

20. Xu D, Mandic DP. 2015 The theory of quaternion matrix derivatives. IEEE Trans. Signal Process. **63**,1543–1556.

21. Ell TA, Sangwine SJ. 2007 Quaternion involutions and anti-involutions. Comput. Math. Appl. **53**, 137–143. ⍰

22. Navarro-Moreno J, Fernández-Alcalá RM, Ruiz-Molina JC. 2014 A quaternion widely linear model for nonlinear Gaussian estimation. IEEE Trans. Signal Process. **62**, 6414–6424.

23. Luna-Elizarrarás ME, Shapiro M. 2011 A survey on the (hyper-) derivatives in complex, quaternionic and Clifford analysis. Milan J. Math. **79**, 521–542.

24. Fueter R. 1934/5 Die Funktionentheorie der Differentialgleichungen $\Delta u=0$ und $\Delta\Delta u=0$ mit vier reellen Variablen. Comment. Math. Helv. **7**, 307–330.

25. Fueter R. 1935/6 Über die analytische Darstellung der regulären Funktionen einer Quaternionenvariablen.Math. Helv. **8**, 371–378.

26. Gentili G, Struppa DC. 2007 A new theory of regular functions of a quaternionic variable. Adv. Math. **216**,279–301.

27. Colombo F, Sabadini I, Struppa DC. 2009 Slice monogenic functions. Israel J. Math. **171**, 385–403.

28. Ujang BC, Took CC, Mandic DP. 2011 Quaternion valued nonlinear adaptive filtering. IEEE Trans. Neural Netw.**22**, 1193–1206.

29. Colombo F, Sabadini I, Struppa DC. 2011 *Noncommutative functional calculus: theory and applications of slice hyperholomorphic functions*. Basel, Switzerland: Birkhäuser.

30. Gentili G, Stoppato C, Struppa DC. 2013 *Regular functions of a quaternionic variable*. Heidelberg, Germany:Springer.

31. Wirtinger W. 1927 Zur formalen theorie der funktionen von mehr komplexen veränderlichen. Math. Ann. **97**,357–375.

32. Vía J, Ramírez D, Santamaría I. 2010 Properness and widely linear processing of quaternion random vectors.IEEE Trans. Inform. Theory **56**, 3502–3515.

33. Isokawa T, Nishimura H, Matsui N. 2012 Quaternionic multilayer perceptron with local analyticity. Information **3**,756–770.

34. Apostol TM 1967 *Calculus*, **vol. I**, 2nd edn. London, UK: John Wiley and Sons Ltd.

35. Took CC, Mandic DP. 2010 A quaternion widely linear adaptive filter. IEEE Trans. Signal Process. **58**,4427–4431.

36. Xu D, Xia Y, Mandic DP. In press. Optimization in quaternion dynamic systems: gradient, Hessian, and learning algorithms. IEEE Trans. Neural Netw. Learn. Syst.

37. Ujang BC, Took CC, Mandic DP. 2010 Split quaternion nonlinear adaptive filtering. Neural Netw. **23**, 426–434.

38. Goh SL, Mandic DP. 2007 An augmented CRTRL for complex-valued recurrent neural networks. Neural Netw.**20**, 1061–1066.

39. Widrow B, McCool J, Ball M . 1975 The complex LMS algorithm. Proc. IEEE **63**, 719–720.

40. Jahanchahi C, Mandic DP . 2013 A unifying approach to quaternion adaptive filtering: addressing the gradient and convergence. (http://arxiv.org/abs/1310.5612)

2

HYPERHOLOMORPHIC FUNCTIONS AND HYPER-CONJUGATE HARMONIC FUNCTIONS OF OCTONION VARIABLES

Su Jin Lim and Kwang Ho Shon*

Department of Mathematics, Pusan National University,
Busan, 609-735, Korea

ABSTRACT

We represent hyperholomorphic functions on octonionic function theory and octonionic differential operators. We research the properties of hyperholomorphic functions, hyper-conjugate harmonic functions and the integral calculus of hyperholomorphic functions by octonion forms.

KEYWORDS

hyperholomorphic function, Clifford analysis, quaternion octonion, pseudoconvex domain

1 INTRODUCTION

The octonions in Clifford algebra are a normed division algebra with eight dimensions over the real numbers larger than the quaternions. The octonions are non-commutative and non-associative but satisfy a weaker form of associativity. The octonions were discovered in 1843 by John T. Graves and constructed in 1845 by A. Cayley. They are referred to as Cayley numbers or the Cayley algebra. The octonions have been applied in fields such as string theory, special theory of relativity and quantum theory. Dentoni and Sce [1] gave a definition of octonionic regular functions and several properties of octonionic regular functions in 1973.

In 2004 and 2006, Kajiwara, Li and Shon [2, 3] obtained some results for the regeneration in complex, quaternion and Clifford analysis, and for the inhomogeneous Cauchy-Riemann system of quaternion and Clifford analysis in ellipsoid.

In 2011, Koriyama and Nôno [4] gave three regularities (*HK*-holomorphy, *HF*-holomorphy, H2-holomorphy) of octonionic functions based on holomorphic mappings in a domain in C^4. Naser [5] and Nôno [6, 7] gave some properties of quaternionic hyperholomorphic functions. For any complex harmonic function f1 in a domain of holomorphy *D* in C^2, we can find a function f2such that f1+f2j will be a function hyperholomorphic in *D* and the Cauchy theorem of hyperholomorphic functions in quaternion analysis. The aim of this paper is to define hyperholomorphic functions with octonion variables in C^4 and investigate the properties of the hyperholomorphic functions of octonion variables. We give the condition of harmonicity in C^4. Then for any complex-valued functions g1(z) and g2(z) satisfying the condition of harmonicity in a pseudoconvex domain Ω in C4, we can find hyper-conjugate harmonic functions g3(z) and g4(z), respectively, on Ω such that g(z)=g1(z)+g2(z)e2+g3(z)e4+g4(z)e6 is a hyperholomorphic function on Ω. Also, we investigate the Cauchy theorem of hyperholomorphic functions in octonion analysis.

2 PRELIMINARIES

The field O≅C4 of octonions

$$z = x_0 + \sum_{i=0}^{7} e_i x_i, \quad x_i(i = 0, \ldots, 7) \in \mathbf{R} \tag{1}$$

is an eight-dimensional non-commutative and non-associative **R**-field generated by eight base elements e0, e1, e2, e3, e4, e5, e6and e7 with the following non-commutative multiplication rules:

$$e_i^2 = -1, \quad e_i e_j = -e_j e_i, \quad e_i e_j e_k = e_i(e_j e_k) \quad (i \neq j \neq k, i \neq 0, j \neq 0, k \neq 0),$$

$$e_1 e_2 = e_3, \quad e_3 e_5 = e_6, \quad e_6 e_7 = e_1, \quad e_1 e_4 = e_5,$$

$$e_5 e_7 = e_2, \quad e_2 e_6 = e_4, \quad e_4 e_7 = e_3.$$

The element e0e_0 is the identities of O and e1 identifies the imaginary unit$\sqrt{-1}$ in the **C**-field of complex numbers. An octonion zgiven by (1) is regarded as z=z1+z2e2+z3e4+z4e6∈O, where z1:=x0+e1x1, z2:=x2+e1x3, z3:=x4+e1x5and z4:=x6+e1x7 are complex numbers in **C**. Thus, we identify O with C4

For the equation z3+8=0 in the complex plane **C**, the three solutions are −2, $1 + \sqrt{3}i, 1 - \sqrt{3}i$ in **C**.

In the octonion O, the equation has solutions whose forms are as follows:

$$z = a + be_1 + ce_2 + de_3 + ee_4 + fe_5 + ge_6 + he_7 \quad (a, b, c, d, e, f, g, h \in \mathbf{R}).$$

Then the equation satisfies $z^3 = (a^3 - 3ab^2 - 3ac^2 - 3ad^2 - 3ae^2 - 3af^2 - 3ag^2 - 3ah^2) + (3a^2b - b^3 - bc^2 - bd^2 - be^2 - bf^2 - bg^2 - bh^2)e_1 + (3a^2c - b^2c - c^3 - cd^2 - ce^2 - cf^2 - cg^2 - ch^2)e_2 + (3a^2d - b^2d - c^2d - d^3 - de^2 - df^2 - dg^2 - dh^2)e_3 + (3a^2e - b^2e - c^2e - d^2e - e^3 - ef^2 - eg^2 - eh^2)e_4 + (3a^2f - b^2f - c^2f - d^2f - e^2f - f^3 - fg^2 - fh^2)e_5 + (3a^2g - b^2g - c^2g - d^2g - e^2g - f^2g - g^3 - gh^2)e_6 + (3a^2h - b^2h - c^2h - d^2h - e^2h - f^2h - g^2h - h^3)e_7$. That is, $a^3 - 3ab^2 - 3ac^2 - 3ad^2 - 3ae^2 - 3af^2 - 3ag^2 - 3ah^2 = -8$, $3a^2b - b^3 - bc^2 - bd^2 - be^2 - bf^2 - bg^2 - bh^2 = 0$, $3a^2c - b^2c - c^3 - cd^2 - ce^2 - cf^2 - cg^2 - ch^2 = 0$, $3a^2d - b^2d - c^2d - d^3 - de^2 - df^2 - dg^2 - dh^2 = 0$, $3a^2e - b^2e - c^2e - d^2e - e^3 - ef^2 - eg^2 - eh^2 = 0$, $3a^2f - b^2f - c^2f - d^2f - e^2f - f^3 - fg^2 - fh^2 = 0$, $3a^2g - b^2g - c^2g - d^2g - e^2g - f^2g - g^3 - gh^2 = 0$, $3a^2h - b^2h - c^2h - d^2h - e^2h - f^2h - g^2h - h^3 = 0$. This means that the equation has infinitely many solutions

$$z = 1 + be_1 + ce_2 + de_3 + ee_4 + fe_5 + ge_6 + he_7$$

$$b^2 + c^2 + d^2 + e^2 + f^2 + g^2 + h^2 = 3 \text{ in } \mathbb{O}.$$

For two octonions $z = \sum_{i=0}^{7} e_i x_i$ and $w = \sum_{i=0}^{7} e_i y_i$, the inner product (z,w)(z, w) is defined as follows:

$$(z, w) := \sum_{i=0}^{7} x_i y_i.$$

Also, the octonionic conjugation z∗z^*, the absolute value |z|$|z|$ of z and an inverse z−1z^{-1} of z in O$\mathbb{O}$ are defined, respectively, by

$$z^* = x_0 - \sum_{i=1}^{7} e_i x_i, \qquad |z| = \sqrt{|z_1|^2 + |z_2|^2 + |z_3|^2 + |z_4|^2}, \qquad z^{-1} = \frac{z^*}{|z|^2} \quad (z \neq 0).$$

Thus, the octonion z∈O$z \in \mathbb{O}$ and the octonion conjugation z∗∈O$z^* \in \mathbb{O}$ have the following forms:

$$z = x_0 + e_1x_1 + e_2x_2 + e_3x_3 + e_4x_4 + e_5x_5 + e_6x_6 + e_7x_7 = z_1 + z_2e_2 + z_3e_4 + z_4e_6$$

and

$$z^* = x_0 - e_1x_1 - e_2x_2 - e_3x_3 - e_4x_4 - e_5x_5 - e_6x_6 - e_7x_7 = \overline{z_1} - z_2e_2 - z_3e_4 - z_4e_6,$$

where z1=x0+e1x1, z2=x2+e1x3, z3=x4+e1x5 and z4=x6+e1x7.

We use the following differential operators:

$$D_\alpha = \frac{\partial}{\partial z_1} - e_2 \frac{\partial}{\partial \overline{z_2}}, \qquad D_\alpha^{*} = \frac{\partial}{\partial \overline{z_1}} + e_2 \frac{\partial}{\partial \overline{z_2}},$$

$$D_\beta = \frac{\partial}{\partial z_3} - e_2 \frac{\partial}{\partial \overline{z_4}}, \qquad D_\beta^{*} = \frac{\partial}{\partial \overline{z_3}} + e_2 \frac{\partial}{\partial \overline{z_4}},$$

where $\partial/\partial z_j,\ \partial/\partial \overline{z_j}\ (j = 1,2,3,4)$ are usual differential operators used in complex analysis. And we use the following octonionic differential operators:

$$D := D_\alpha - e_4 D_\beta^{*}, \qquad D^\star := D_\alpha^{*} + e_4 D_\beta^{*}.$$

The operator

$$DD^\star = \frac{\partial^2}{\partial z_1 \partial \overline{z_1}} + \frac{\partial^2}{\partial z_2 \partial \overline{z_2}} + \frac{\partial^2}{\partial z_3 \partial \overline{z_3}} + \frac{\partial^2}{\partial z_4 \partial \overline{z_4}}$$
$$= \frac{1}{4}\left(\frac{\partial^2}{\partial x_0^2} + \frac{\partial^2}{\partial x_1^2} + \frac{\partial^2}{\partial x_2^2} + \frac{\partial^2}{\partial x_3^2} + \frac{\partial^2}{\partial x_4^2} + \frac{\partial^2}{\partial x_5^2} + \frac{\partial^2}{\partial x_6^2} + \frac{\partial^2}{\partial x_7^2} \right)$$

is the usual complex Laplacian Δ.

3 SOME PROPERTIES OF HYPERHOLOMORPHIC FUNCTIONS ON O$\mathcal{O}$

Let Ω be an open set in C4$\mathbb{C}^4$. The function g(z)$g(z)$ is defined by the following form in Ω with value in O$\mathcal{O}$:

$$g(z) = g_1(z) + g_2(z)e_2 + g_3(z)e_4 + g_4(z)e_6 = \{g_1(z) + g_3(z)e_4\} + \{g_2(z) + g_4(z)e_4\}e_2,$$

where z=(z1,z2,z3,z4) and g1(z), g2(z), g3(z) and g4(z) are complex-valued functions.

Definition 3.1 Let Ω be an open set in C4C^4. A function g(z)$g(z)$ is said to be L(R)-hyperholomorphic in Ω if the following two conditions are satisfied:

*(a)*gk(z) (k=1,2,3,4) are continuously differential functions in Ω, and

(b) $D^{\star}g = 0 \qquad (gD^{\star} = 0) \quad \text{in } \Omega.$

When we deal with an L-hyperholomorphic function g(z) in Ω⊂C4, for simplicity, we often say that g(z)$g(z)$ is a hyperholomorphic function in Ω⊂C4.

Equation (2) is applied to g(z) as follows:

$$
\begin{aligned}
D^{\star}g &= \left(D_{\alpha}^{*} + e_4 D_{\beta}^{*}\right)\left(g_1(z) + g_2(z)e_2 + g_3(z)e_4 + g_4(z)e_6\right) \\
&= \left(D_{\alpha}^{*}g_1 - D_{\beta}\overline{g_3}\right) + \left(D_{\alpha}^{*}g_2 - D_{\beta}\overline{g_4}\right)e_2 \\
&\quad + \left(D_{\alpha}^{*}g_3 + D_{\beta}\overline{g_1}\right)e_4 + \left(D_{\alpha}^{*}g_4 + D_{\beta}\overline{g_2}\right)e_6.
\end{aligned}
$$

If the following equations:

$$
D_{\alpha}^{*}g_1 = D_{\beta}\overline{g_3}, \qquad D_{\alpha}^{*}g_2 = D_{\beta}\overline{g_4}, \qquad D_{\alpha}^{*}g_3 = -D_{\beta}\overline{g_1}, \qquad D_{\alpha}^{*}g_4 = -D_{\beta}\overline{g_2} \tag{3}
$$

are satisfied, the function g(z)$g(z)$ is a hyperholomorphic function in Ω. The equations in (3) are the corresponding o-Cauchy-Riemann equations in C4C^4.

Remark 3.2 We redefine equations (3) as follows:

$$
\begin{aligned}
&\frac{\partial g_1}{\partial \overline{z_1}} = \frac{\partial \overline{g_3}}{\partial z_3}, \quad \frac{\partial g_1}{\partial \overline{z_2}} = -\frac{\partial \overline{g_3}}{\partial \overline{z_4}}, \quad \frac{\partial g_3}{\partial \overline{z_1}} = -\frac{\partial \overline{g_1}}{\partial z_3}, \quad \frac{\partial g_3}{\partial \overline{z_2}} = \frac{\partial \overline{g_1}}{\partial \overline{z_4}}, \\
&\frac{\partial g_2}{\partial \overline{z_1}} = \frac{\partial \overline{g_4}}{\partial z_3}, \quad \frac{\partial g_2}{\partial \overline{z_2}} = -\frac{\partial \overline{g_4}}{\partial \overline{z_4}}, \quad \frac{\partial g_4}{\partial \overline{z_1}} = -\frac{\partial \overline{g_2}}{\partial z_3}, \quad \frac{\partial g_4}{\partial \overline{z_2}} = \frac{\partial \overline{g_2}}{\partial \overline{z_4}}.
\end{aligned} \tag{4}
$$

We call that equations (4) are the condition of harmonicity.

Remark 3.3 We redefine equations (4) in R8$\mathbb{R}^8$ as follows:

$$\frac{\partial u_0}{\partial x_0} - \frac{\partial u_1}{\partial x_1} - \frac{\partial u_4}{\partial x_4} + \frac{\partial u_5}{\partial x_5} = 0, \qquad \frac{\partial u_1}{\partial x_0} + \frac{\partial u_0}{\partial x_1} + \frac{\partial u_5}{\partial x_4} + \frac{\partial u_4}{\partial x_5} = 0,$$

$$\frac{\partial u_0}{\partial x_2} - \frac{\partial u_1}{\partial x_3} + \frac{\partial u_4}{\partial x_6} + \frac{\partial u_5}{\partial x_7} = 0, \qquad \frac{\partial u_1}{\partial x_2} + \frac{\partial u_0}{\partial x_3} - \frac{\partial u_5}{\partial x_6} + \frac{\partial u_4}{\partial x_7} = 0,$$

$$\frac{\partial u_2}{\partial x_0} - \frac{\partial u_3}{\partial x_1} - \frac{\partial u_6}{\partial x_4} + \frac{\partial u_7}{\partial x_5} = 0, \qquad \frac{\partial u_3}{\partial x_0} + \frac{\partial u_2}{\partial x_1} + \frac{\partial u_7}{\partial x_4} + \frac{\partial u_6}{\partial x_5} = 0,$$

$$\frac{\partial u_2}{\partial x_2} - \frac{\partial u_3}{\partial x_3} + \frac{\partial u_6}{\partial x_6} + \frac{\partial u_7}{\partial x_7} = 0, \qquad \frac{\partial u_3}{\partial x_2} + \frac{\partial u_2}{\partial x_3} - \frac{\partial u_7}{\partial x_6} + \frac{\partial u_6}{\partial x_7} = 0,$$

$$\frac{\partial u_4}{\partial x_0} - \frac{\partial u_5}{\partial x_1} + \frac{\partial u_0}{\partial x_4} - \frac{\partial u_1}{\partial x_5} = 0, \qquad \frac{\partial u_5}{\partial x_0} + \frac{\partial u_4}{\partial x_1} - \frac{\partial u_1}{\partial x_4} - \frac{\partial u_0}{\partial x_5} = 0,$$

$$\frac{\partial u_4}{\partial x_2} - \frac{\partial u_5}{\partial x_3} - \frac{\partial u_0}{\partial x_6} - \frac{\partial u_1}{\partial x_7} = 0, \qquad \frac{\partial u_5}{\partial x_2} + \frac{\partial u_4}{\partial x_3} + \frac{\partial u_1}{\partial x_6} - \frac{\partial u_0}{\partial x_7} = 0,$$

$$\frac{\partial u_6}{\partial x_0} - \frac{\partial u_7}{\partial x_1} + \frac{\partial u_2}{\partial x_4} - \frac{\partial u_3}{\partial x_5} = 0, \qquad \frac{\partial u_7}{\partial x_0} + \frac{\partial u_6}{\partial x_1} - \frac{\partial u_3}{\partial x_4} - \frac{\partial u_2}{\partial x_5} = 0,$$

$$\frac{\partial u_6}{\partial x_2} - \frac{\partial u_7}{\partial x_3} - \frac{\partial u_2}{\partial x_6} - \frac{\partial u_3}{\partial x_7} = 0, \qquad \frac{\partial u_7}{\partial x_2} + \frac{\partial u_6}{\partial x_3} + \frac{\partial u_3}{\partial x_6} - \frac{\partial u_2}{\partial x_7} = 0,$$

where g1=u0+e1u1, g2=u2+e1u3, g3=u4+e1u5 and g4=u6+e1u7 for real-valued functions ui (i=0,...,7).

Lemma 3.4

(i)If the function g(z) *is hyperholomorphic in an open set* Ω *in* C4, *then the functions* g1(z), g2(z), g3(z) *and* g4(z) *are of class* C∞ *in* Ω.

(ii)If the function g(z) *satisfies the condition of harmonicity in an open set* Ω *in* C4, *then the functions* g1(z), g2(z), g3(z) *and* g4(z) *are harmonic in* Ω.

Proof We have

$$DD^{*}g_1 = \frac{\partial^2 g_1}{\partial z_1 \partial \overline{z_1}} + \frac{\partial^2 g_1}{\partial z_2 \partial \overline{z_2}} + \frac{\partial^2 g_1}{\partial z_3 \partial \overline{z_3}} + \frac{\partial^2 g_1}{\partial z_4 \partial \overline{z_4}}$$
$$= \frac{\partial}{\partial z_1}\left(\frac{\partial \overline{g_3}}{\partial z_3}\right) + \frac{\partial}{\partial z_2}\left(-\frac{\partial \overline{g_3}}{\partial \overline{z_4}}\right) + \frac{\partial}{\partial z_3}\left(-\frac{\partial \overline{g_3}}{\partial z_1}\right) + \frac{\partial}{\partial \overline{z_4}}\left(\frac{\partial \overline{g_3}}{\partial z_2}\right)$$
$$= 0,$$

and the functions g2, g3 and g4 are proved by a similar method as in the proof of the case of g1. And, by (i), gj(z) (j=1,2,3,4) are of class C∞ functions in Ω. □

Definition 3.5 Let Ω⊂Cn be an open set with a C2C^2 boundary. Let Ω={z;ρ(z)<0}, where *ρ* is in C2 in a neighborhood ofΩ¯ and grad ρ≠0 on *b* Ω. Then Ω is pseudoconvex if

$$\sum_{j,k=1}^{n} \frac{\partial^2 \rho}{\partial z_j \partial \overline{z_k}}(z) w_j \overline{w_k} \geq 0,$$

for all z∈bΩ and w∈Cn satisfying $\sum_{j=1}^{n} \frac{\partial \rho}{\partial z_j}(z) w_j = 0.$.

Consider an automorphism *γ*:

$$(z_1, z_2, z_3, s_4) = \gamma(z_1, z_2, z_3, z_4) := (z_1, z_2, z_3, \overline{z_4})$$

of C4C^4. A domain Ω in C4≅O is said to be pseudoconvex with respect to the complex variables z1,z2,z3,z4¯, if γ(Ω) is a pseudoconvex domain of the space C4 of four complex variables z1,z2,z3,s4 in the sense of complex analysis.

Theorem 3.6 *Let* Ω *be a domain in* C4≅O, *which is a pseudoconvex domain with respect to the complex variables* z1,z2,z3,z4¯*and let* g1(z) *and* g2(z) *be complex-valued functions of class* C2 *on* Ω *satisfying the condition of harmonicity* (4).*Then there exist hyper-conjugate harmonic functions* g3(z) *and* g4(z), *respectively, of class* C2 *on* Ω *such that* g(z)$g(z)$ *is a hyperholomorphic function on* Ω.

Proof We consider the 1-forms and the differential operator on $\gamma(\Omega)$:

$$\psi_1 := -\frac{\partial \overline{g_1}}{\partial z_3} d\overline{z_1} + \frac{\partial \overline{g_1}}{\partial \overline{z_4}} d\overline{z_2} + \frac{\partial \overline{g_1}}{\partial z_1} d\overline{z_3} - \frac{\partial \overline{g_1}}{\partial z_2} dz_4,$$

$$\psi_2 := -\frac{\partial \overline{g_2}}{\partial z_3} d\overline{z_1} + \frac{\partial \overline{g_2}}{\partial \overline{z_4}} d\overline{z_2} + \frac{\partial \overline{g_2}}{\partial z_1} d\overline{z_3} - \frac{\partial \overline{g_2}}{\partial z_2} dz_4,$$

and

$$\delta = \frac{\partial}{\partial \overline{z_1}} d\overline{z_1} + \frac{\partial}{\partial \overline{z_2}} d\overline{z_2} + \frac{\partial}{\partial \overline{z_3}} d\overline{z_3} + \frac{\partial}{\partial z_4} dz_4.$$

We operate the operator δ from the left-hand side of the 1-forms ψ1ψ_1 and ψ2ψ_2 on $\gamma(\Omega)$:

$$\begin{aligned}\delta\psi_1 = &\left(\frac{\partial^2 \overline{g_1}}{\partial \overline{z_1}\,\partial \overline{z_4}} + \frac{\partial^2 \overline{g_1}}{\partial \overline{z_2}\,\partial z_3}\right) d\overline{z_1} \wedge d\overline{z_2} + \left(\frac{\partial^2 \overline{g_1}}{\partial \overline{z_1}\,\partial z_1} + \frac{\partial^2 \overline{g_1}}{\partial \overline{z_3}\,\partial z_3}\right) d\overline{z_1} \wedge d\overline{z_3}\\ &+ \left(-\frac{\partial^2 \overline{g_1}}{\partial \overline{z_1}\,\partial z_2} + \frac{\partial^2 \overline{g_1}}{\partial z_4\,\partial z_3}\right) d\overline{z_1} \wedge dz_4 + \left(\frac{\partial^2 \overline{g_1}}{\partial \overline{z_2}\,\partial z_1} - \frac{\partial^2 \overline{g_1}}{\partial \overline{z_3}\,\partial \overline{z_4}}\right) d\overline{z_2} \wedge d\overline{z_3}\\ &+ \left(-\frac{\partial^2 \overline{g_1}}{\partial \overline{z_2}\,\partial z_2} - \frac{\partial^2 \overline{g_1}}{\partial z_4\,\partial \overline{z_4}}\right) d\overline{z_2} \wedge dz_4 + \left(-\frac{\partial^2 \overline{g_1}}{\partial \overline{z_3}\,\partial z_2} - \frac{\partial^2 \overline{g_1}}{\partial z_4\,\partial z_1}\right) d\overline{z_3} \wedge dz_4,\end{aligned}$$

and

$$\begin{aligned}\delta\psi_2 = &\left(\frac{\partial^2 \overline{g_2}}{\partial \overline{z_1}\,\partial \overline{z_4}} + \frac{\partial^2 \overline{g_2}}{\partial \overline{z_2}\,\partial z_3}\right) d\overline{z_1} \wedge d\overline{z_2} + \left(\frac{\partial^2 \overline{g_2}}{\partial \overline{z_1}\,\partial z_1} + \frac{\partial^2 \overline{g_2}}{\partial \overline{z_3}\,\partial z_3}\right) d\overline{z_1} \wedge d\overline{z_3}\\ &+ \left(-\frac{\partial^2 \overline{g_2}}{\partial \overline{z_1}\,\partial z_2} + \frac{\partial^2 \overline{g_2}}{\partial z_4\,\partial z_3}\right) d\overline{z_1} \wedge dz_4 + \left(\frac{\partial^2 \overline{g_2}}{\partial \overline{z_2}\,\partial z_1} - \frac{\partial^2 \overline{g_2}}{\partial \overline{z_3}\,\partial \overline{z_4}}\right) d\overline{z_2} \wedge d\overline{z_3}\\ &+ \left(-\frac{\partial^2 \overline{g_2}}{\partial \overline{z_2}\,\partial z_2} - \frac{\partial^2 \overline{g_2}}{\partial z_4\,\partial \overline{z_4}}\right) d\overline{z_2} \wedge dz_4 + \left(-\frac{\partial^2 \overline{g_2}}{\partial \overline{z_3}\,\partial z_2} - \frac{\partial^2 \overline{g_2}}{\partial z_4\,\partial z_1}\right) d\overline{z_3} \wedge dz_4.\end{aligned}$$

By the condition of harmonicity (4), all coefficients vanish. From Hörmander [8], the δ-closed forms ψ1 and ψ2 of z1,z2,z3,z4are δ-exact forms on γ(Ω)$\gamma(\Omega)$. Since Ω is a pseudoconvex domain, there exist hyper-conjugate harmonic functions g3(z) and g4(z) of class C∞ on Ω with ∂¯-closed forms γ−1ψ1=∂¯g3(z) and γ−1ψ2=∂¯g4(z) on

Ω of z1,z2,z3,s4 are ∂¯∂̄-exact (0,1)(0,1)-forms on Ω such that g(z) is a hyperholomorphic function on Ω (see Krantz [9]). □

Theorem 3.7 *Let* Ω *be a domain in* C4≅O, *which is a pseudoconvex domain with respect to the complex variables* z1,z2,z3,z4¯ *and let* J1(z)=g1(z)+g3(z)e4 *be a complex-valued function of class* C2C^2 *on* Ω *satisfying the condition of harmonicity* (4). *Then there exists a hyper-conjugate harmonic function* J2(z)=g2(z)+g4(z)e4 *of class* C2C^2 *on* Ω *such that* g(z)=J1(z)+J2(z)e2 *is a hyperholomorphic function on* Ω.

Proof We consider the 1-form and the differential operator on γ(Ω):

$$\psi := \left(\frac{\partial\overline{g_4}}{\partial z_3} - \frac{\partial\overline{g_2}}{\partial z_3}e_4\right)d\overline{z_1} + \left(-\frac{\partial\overline{g_4}}{\partial\overline{z_4}} + \frac{\partial\overline{g_2}}{\partial\overline{z_4}}e_4\right)d\overline{z_2}$$
$$+\left(-\frac{\partial\overline{g_4}}{\partial z_1} + \frac{\partial\overline{g_2}}{\partial z_1}e_4\right)d\overline{z_3} + \left(\frac{\partial\overline{g_4}}{\partial z_2} - \frac{\partial\overline{g_2}}{\partial z_2}e_4\right)dz_4$$

and

$$\delta = \frac{\partial}{\partial\overline{z_1}}d\overline{z_1} + \frac{\partial}{\partial\overline{z_2}}d\overline{z_2} + \frac{\partial}{\partial\overline{z_3}}d\overline{z_3} + \frac{\partial}{\partial z_4}dz_4.$$

We operate the operator δ from the left-hand side of the 1-form ψ on γ(Ω):

$$\delta\psi = \left\{\left(-\frac{\partial^2\overline{g_4}}{\partial\overline{z_1}\,\partial\overline{z_4}} - \frac{\partial^2\overline{g_4}}{\partial\overline{z_2}\,\partial z_3}\right) + \left(\frac{\partial^2\overline{g_2}}{\partial\overline{z_1}\,\partial\overline{z_4}} + \frac{\partial^2\overline{g_2}}{\partial\overline{z_2}\,\partial z_3}\right)e_4\right\}d\overline{z_1}\wedge d\overline{z_2}$$
$$+\left\{\left(-\frac{\partial^2\overline{g_4}}{\partial\overline{z_1}\,\partial z_1} - \frac{\partial^2\overline{g_4}}{\partial\overline{z_3}\,\partial z_3}\right) + \left(\frac{\partial^2\overline{g_2}}{\partial\overline{z_1}\,\partial z_1} + \frac{\partial^2\overline{g_2}}{\partial\overline{z_3}\,\partial z_3}\right)e_4\right\}d\overline{z_1}\wedge d\overline{z_3}$$
$$+\left\{\left(\frac{\partial^2\overline{g_4}}{\partial\overline{z_1}\,\partial z_2} - \frac{\partial^2\overline{g_4}}{\partial z_4\,\partial z_3}\right) + \left(-\frac{\partial^2\overline{g_2}}{\partial\overline{z_1}\,\partial z_2} + \frac{\partial^2\overline{g_2}}{\partial z_4\,\partial z_3}\right)e_4\right\}d\overline{z_1}\wedge dz_4$$
$$+\left\{\left(-\frac{\partial^2\overline{g_4}}{\partial\overline{z_2}\,\partial z_1} + \frac{\partial^2\overline{g_4}}{\partial\overline{z_3}\,\partial\overline{z_4}}\right) + \left(\frac{\partial^2\overline{g_2}}{\partial\overline{z_2}\,\partial z_1} - \frac{\partial^2\overline{g_2}}{\partial\overline{z_3}\,\partial\overline{z_4}}\right)e_4\right\}d\overline{z_2}\wedge d\overline{z_3}$$
$$+\left\{\left(\frac{\partial^2\overline{g_4}}{\partial\overline{z_2}\,\partial z_2} + \frac{\partial^2\overline{g_4}}{\partial z_4\,\partial\overline{z_4}}\right) + \left(-\frac{\partial^2\overline{g_2}}{\partial\overline{z_2}\,\partial z_2} - \frac{\partial^2\overline{g_2}}{\partial z_4\,\partial\overline{z_4}}\right)e_4\right\}d\overline{z_2}\wedge dz_4$$
$$+\left\{\left(\frac{\partial^2\overline{g_4}}{\partial\overline{z_3}\,\partial z_2} + \frac{\partial^2\overline{g_4}}{\partial z_4\,\partial z_1}\right) + \left(-\frac{\partial^2\overline{g_2}}{\partial\overline{z_3}\,\partial z_2} - \frac{\partial^2\overline{g_2}}{\partial z_4\,\partial z_1}\right)e_4\right\}d\overline{z_3}\wedge dz_4.$$

By the same method as the proof of Theorem 3.6, our result is proved. □

Theorem 3.8 *Let* g(z)$g(z)$ *be a hyperholomorphic function in a domain G of* O$\mathbb{O}$ *and let*

$$\begin{aligned}\tau &= dz_1 \wedge dz_2 \wedge dz_3 \wedge dz_4 \wedge d\overline{z_2} \wedge d\overline{z_3} \wedge d\overline{z_4}\\ &+ dz_1 \wedge dz_2 \wedge dz_3 \wedge dz_4 \wedge d\overline{z_1} \wedge d\overline{z_3} \wedge d\overline{z_4}\\ &- dz_1 \wedge dz_2 \wedge dz_4 \wedge d\overline{z_1} \wedge d\overline{z_2} \wedge d\overline{z_3} \wedge d\overline{z_4} e_4\\ &- dz_1 \wedge dz_2 \wedge dz_3 \wedge dz_4 \wedge d\overline{z_1} \wedge d\overline{z_2} \wedge d\overline{z_3} e_4.\end{aligned}$$

Then for any domain Ω⊂G$\Omega \subset G$ *with smooth boundary b* Ω,

$$\int_{b\Omega} \tau g = 0, \tag{5}$$

where τg *is the octonion product of the form* τ *on the function* g(z)$g(z)$.

Proof Let

$$\begin{aligned}\tau_{(1)} &= dz_1 \wedge dz_2 \wedge dz_3 \wedge dz_4 \wedge d\overline{z_2} \wedge d\overline{z_3} \wedge d\overline{z_4},\\ \tau_{(2)} &= dz_1 \wedge dz_2 \wedge dz_3 \wedge dz_4 \wedge d\overline{z_1} \wedge d\overline{z_3} \wedge d\overline{z_4},\\ \tau_{(3)} &= dz_1 \wedge dz_2 \wedge dz_4 \wedge d\overline{z_1} \wedge d\overline{z_2} \wedge d\overline{z_3} \wedge d\overline{z_4},\\ \tau_{(4)} &= dz_1 \wedge dz_2 \wedge dz_3 \wedge dz_4 \wedge d\overline{z_1} \wedge d\overline{z_2} \wedge d\overline{z_3}.\end{aligned}$$

By the rule of octonion multiplications,

$$\begin{aligned}\tau g &= (\tau_{(1)} + \tau_{(2)} - \tau_{(3)} e_4 - \tau_{(4)} e_4)(g_1 + g_2 e_2 + g_3 e_4 + g_4 e_6)\\ &= g_1 \tau_{(1)} + g_1 \tau_{(2)} - \overline{g_1} \tau_{(3)} e_4 - \overline{g_1} \tau_{(4)} e_4\\ &\quad + g_2 \tau_{(1)} e_2 + g_2 \tau_{(2)} e_2 - \overline{g_2} \tau_{(3)} e_6 - \overline{g_2} \tau_{(4)} e_6\end{aligned}$$

$$+ g_3 \tau_{(1)} e_4 + g_3 \tau_{(2)} e_4 + \overline{g_3} \tau_{(3)} + \overline{g_3} \tau_{(4)}$$

$$+ g_4 \tau_{(1)} e_6 + g_4 \tau_{(2)} e_6 + \overline{g_4} \tau_{(3)} e_2 + \overline{g_4} \tau_{(4)} e_2.$$

Hence,

$$d(\tau g) = \left(-\frac{\partial g_1}{\partial \overline{z_1}} - \frac{\partial g_1}{\partial \overline{z_2}} + \frac{\partial \overline{g_3}}{\partial z_3} - \frac{\partial \overline{g_3}}{\partial \overline{z_4}}\right) dV + \left(-\frac{\partial g_2}{\partial \overline{z_1}} - \frac{\partial g_2}{\partial \overline{z_2}} + \frac{\partial \overline{g_4}}{\partial z_3} - \frac{\partial \overline{g_4}}{\partial \overline{z_4}}\right) dVe_2$$
$$+ \left(-\frac{\partial \overline{g_1}}{\partial z_3} + \frac{\partial \overline{g_1}}{\partial \overline{z_4}} - \frac{\partial g_3}{\partial \overline{z_1}} - \frac{\partial g_3}{\partial \overline{z_2}}\right) dVe_4 + \left(-\frac{\partial \overline{g_2}}{\partial z_3} + \frac{\partial \overline{g_2}}{\partial \overline{z_4}} - \frac{\partial g_4}{\partial \overline{z_1}} - \frac{\partial g_4}{\partial \overline{z_2}}\right) dVe_6,$$

where $dV = dz_1 \wedge dz_2 \wedge dz_3 \wedge dz_4 \wedge d\overline{z_1} \wedge d\overline{z_2} \wedge d\overline{z_3} \wedge d\overline{z_4}$, and by the condition of harmonicity (4), d(τg)=0$d(\tau g) = 0$. By Stoke's theorem, we have

$$\int_{b\Omega} \tau g = \int_{\Omega} d(\tau g) = 0.$$

ACKNOWLEDGEMENTS

The second author was supported by the Basic Science Research Program through the National Research Foundation of Korea (NRF) funded by the Ministry of Education, Science and Technology (2010-0009646), and by the Research Fund Program of Research Institute for Basic Sciences, Pusan National University, Korea, 2012, Project No. RIBS-PNU-2012-101.

COMPETING INTERESTS

The authors declare that they have no competing interests.

AUTHORS' CONTRIBUTIONS

All authors read and approved the final manuscript.

REFERENCES

1. Dentoni P, Sce M: Funzioni regolari nell'algebra di Cayley. *Rend. Sem. Mat. Univ. Padora* 1973, 50: 251–267.
2. Kajiwara J, Li XD, Shon KH: Regeneration in complex, quaternion and Clifford analysis. Complex Analysis and Appl. 9. In *Proc. the 9th International Conf. on Finite or Infinite Dimen*. Kluwer Academic, Dordrecht; 2004:287–298.
3. Kajiwara J, Li XD, Shon KH: Function spaces in complex and Clifford analysis, inhomogeneous Cauchy Riemann system of quaternion and Clifford analysis in ellipsoid. Complex Analysis and Appl. 14. In *Proc. the 14th International Conf. on Finite or Infinite Dimen*. Hue Univ., Vietnam; 2006:127–155.
4. Koriyama H, Nôno K: On regularities of octonionic functions and holomorphic mappings. *Bull. Fukuoka Univ. Edu. Part III* 2011, 60: 11–28.
5. Naser M: Hyperholomorphic functions. *Silberian Math. J.* 1971, 12: 959–968.
6. Nôno K: Hyperholomorphic functions of a quaternion variable. *Bull. Fukuoka Univ. Edu.* 1983, 32: 21–37.
7. Nôno K: Characterization of domains of holomorphy by the existence of hyper-conjugate harmonic functions. *Rev. Roum. Math. Pures Appl.* 1986, 31(2):159–161.
8. Hörmander L: *An Introduction to Complex Analysis in Several Variables*. North-Holland, Amsterdam; 1966.

9. Krantz SG: *Function Theory of Several Complex Variables*. Am. Math. Soc., Providence; 2001.

3

$H\Lambda$-REGULAR VECTOR FUNCTIONS AND THEIR BOUNDARY VALUE PROBLEMS

Piwen Yang* and Dan Li

Department of Mathematics, Sichuan Normal University, Chengdu 610066, P.R. China

ABSTRACT

Let $D = \begin{pmatrix} \lambda + \frac{\partial}{\partial t} & 2\frac{\partial}{\partial \bar{z}} \\ 2\frac{\partial}{\partial z} & \lambda - \frac{\partial}{\partial t} \end{pmatrix}$ where λ is a positive real constant. In this paper, by using the methods from quaternion calculus, we investigate the HλH_λ-regular vector functions, that is, the complex vector solutions $\Psi(t,z) = \begin{pmatrix} \psi_1(t,z) \\ \psi_2(t,z) \end{pmatrix}$ of the equation DΨ=0, and work out a systematic theory analogous to quaternionic regular functions. Differing from that, the component functions of quaternionic regular functions are harmonic, the component functions of HλH_λ-regular functions satisfy the modified Helmholtz equation, that is (λ2−Δ)ψi=0, i=1,2$i = 1,2$. We give out a distribution solution of the inhomogeneous equation Du=f and study some properties of the solution. Moreover, we discuss some boundary value problems for HλH_λ-regular functions and solutions of equation Du=f.

KEYWORDS

quaternion calculus, Hλ-regular vector function, modified Helmholtz equation, Riemann-Hilbert type, boundary value problem

It is well known that the theories of holomorphic functions of one complex variable and regular functions of quaternion as well as Clifford calculus are closely connected with the theory of harmonic functions, *i.e.*, their component functions are all harmonic. But side by side with the Laplace operator is the Helmholtz operator and modified Helmholtz operator

$$\Delta_\lambda = \lambda^2 \pm \Delta, \quad \lambda(\in \mathbb{R}) \neq 0,$$

which play an important role and are often met in application. In recent years, it has been considered that by replacing the harmonic function with the solutions of Helmholtz equation and modified Helmholtz equation, the theory of regular functions is naturally generalized in quaternion calculus and Clifford calculus. The theory has been well developed and has been applied to the research of some partial differential equations such as Helmholtz equation, Klein-Cordon equation, and Schroding equation. The corresponding results can be found in [1–3, 5–11, 13–15].

Let H(R) and H(C) denote the real and complex quaternion space respectively. Their basis elements 1, *i*, *j*, *k* satisfy the following relations: $i^2 = j^2 = k^2 = -1,\ ij = -ji = k,\ jk = -kj = i.$.

In [2, 3], the authors introduced a differential operator of first order $D_\lambda = \lambda + i\frac{\partial}{\partial x_1} + j\frac{\partial}{\partial x_2} + k\frac{\partial}{\partial x_3}$,where λ is a positive real constant. It is easy to see that

$$-D_{-\lambda}D_{\lambda} = \lambda^2 + \Delta,$$

where λ2+Δ$\lambda^2 + \Delta$ is namely the 3-dimensional Helmholtz operator. A quaternion function theory associated with the operator was established which involved the Pompeiu formula corresponding to DλD_λ, the Cauchy integral formula for solutions of equation Dλu=0$D_\lambda u = 0$, the Plemelj formula of Cauchy type integral and the theory of operator TλT_λ. By using these results, the authors investigated the Dirichlet boundary problems for Helmholtz equation

$$(\lambda^2 + \Delta)u = f.$$

Since the operator λ2−Δ can not be factorized into the product of two differential operators of first order in H(R), the quaternion function theory about modified Helmholtz equation was developed in complex quaternion space H(C), namely the operator λ2+Δ, λ∈C and some related equations were directly investigated by H(C). However, different from H(R), H(C) is a Euclidean 8-space; and since there exists a set of zero divisors in H(C), a non-zero complex quaternion is not necessarily invertible. There exist many differences between the two theories.

In this article, we shall use the quasi-quaternion space introduced in [18, 19] and transform the modified Helmholtz operator into matric form $(\lambda^2-\Delta)e0=DD'$. By using the quaternion technique, we obtain a systematic theory about the Hλ-regular vector functions, that is, the complex vector solutions $\Psi(t,z) = \binom{\psi_1(t,z)}{\psi_2(t,z)}$ of the equation DΨ=0, analogous to the quaternion regular function. Because the Hλ-regular vector functions are two-dimensional complex vector functions, this is more similar to the case of H(R)

For applications of partial differential equations, the research of boundary value problems is very important. How should appropriate boundary data be chosen for the Helmholtz equation or modified Helmholtz equation of first order? So far, there have been very few research works on the aspect. In this article, we introduce and investigate some Riemann-Hilbert type boundary value problems for HλH_λ-regular vector functions and solutions of the equation Du=f$Du=f$, obtain general solutions and solvable conditions respectively in different cases.

1 SOME NOTATIONS AND DEFINITIONS

Denote

$$e_0 = \begin{pmatrix} 1 & 0 \\ 0 & 1 \end{pmatrix}, \quad e_1 = \begin{pmatrix} 1 & 0 \\ 0 & -1 \end{pmatrix}, \quad e_2 = \begin{pmatrix} 0 & 1 \\ 1 & 0 \end{pmatrix}, \quad e_3 = \begin{pmatrix} 0 & i \\ -i & 0 \end{pmatrix}.$$

It is easy to see that

$$e_1^2 = e_2^2 = e_3^2 = e_0, \qquad e_1e_2 = -e_2e_1 = -ie_3,$$

$$e_2e_3 = -e_3e_2 = -ie_1, \qquad e_3e_1 = -e_1e_3 = -ie_2.$$

Henceforth we shall abbreviate e0e_0 to 1.

Introduce the three-dimensional modified Helmholtz operator D=λ+∇ of first order, where $\nabla = \frac{\partial}{\partial t}e_1 + \frac{\partial}{\partial x}e_2 + \frac{\partial}{\partial y}e_3$, λ is a positive real constant. Define D′=λ−∇, then DD′=D′D=(λ2−∆)e0, where ∆ is the three-dimensional Laplace operator. The matrix forms of *D*, D′ are

$$D = \begin{pmatrix} \lambda + \frac{\partial}{\partial t} & 2\frac{\partial}{\partial \bar{z}} \\ 2\frac{\partial}{\partial z} & \lambda - \frac{\partial}{\partial t} \end{pmatrix}, \qquad D' = \begin{pmatrix} \lambda - \frac{\partial}{\partial t} & -2\frac{\partial}{\partial \bar{z}} \\ -2\frac{\partial}{\partial z} & \lambda + \frac{\partial}{\partial t} \end{pmatrix},$$

where

$$\frac{\partial}{\partial \bar{z}} = \frac{1}{2}\left(\frac{\partial}{\partial x} + i\frac{\partial}{\partial y}\right),$$

$$\frac{\partial}{\partial z} = \frac{1}{2}\left(\frac{\partial}{\partial x} - i\frac{\partial}{\partial y}\right),$$

and then

$$DD' = D'D = \begin{pmatrix} \lambda^2 - \Delta & 0 \\ 0 & \lambda^2 - \Delta \end{pmatrix}.$$

Let Ω be a region in R×C which identifies with R3. $\Psi(t,z) = \begin{pmatrix} \psi_1(t,z) \\ \psi_2(t,z) \end{pmatrix}$ is a complex vector function defined in Ω. If Ψ(t,z)∈C1(Ω) and satisfies the equation

$$D\Psi = 0, \tag{1}$$

then Ψ(t,z) will be called Hλ-regular vector function in Ω.

2 POMPEIU FORMULA AND CAUCHY INTEGRAL FORMULA OF HΛH_λ-REGULAR VECTOR FUNCTION

Let Ω be a bounded domain in R×C with piecewise smooth boundary *S*. U(t,z), V(t,z) are two-dimensional complex vector functions defined in Ω and U(t,z),V(t,z)∈C1(Ω)∩C(Ω¯¯¯). By the divergence theorem

$$\begin{aligned}\int_\Omega [(U\nabla)V + U(\nabla V)]\,d\sigma &= \int_\Omega \left[\frac{\partial}{\partial x_1}(Ue_1V) + \frac{\partial}{\partial x_2}(Ue_2V) + \frac{\partial}{\partial x_3}(Ue_3V)\right] d\sigma \\ &= \int_S U\Im V\,dS,\end{aligned} \tag{2}$$

where I=e1cosα1+e2cosα2+e3cosα3, (cosα1,cosα2,cosα3) denotes the unit outward normal to the surface *S*. From the equality (2), we have

$$\int_{\Omega}[U(DV)-(UD')V]\,d\sigma = \int_S U\Im V\,dS. \tag{3}$$

It is easy to show that $\frac{1}{4\pi r}e^{-\lambda r}, r=(t^2+x^2+y^2)^{\frac{1}{2}}$, is a fundamental solution of the modified Helmholtz operator λ2−Δ. When r≠0, $\lambda^2-\Delta$. When $r\neq 0$, $(\lambda^2-\Delta)(\frac{1}{4\pi r}e^{-\lambda r})=0$. We write

$$E(t,z)=D\left(\frac{1}{4\pi r}e^{-\lambda r}\right)=\frac{1}{4\pi}\left[\frac{\lambda}{r}-\left(\frac{\lambda}{r^2}+\frac{1}{r^3}\right)(te_1+xe_2+ye_3)\right]e^{-\lambda r}.$$

Suppose u(t,z) is a complex vector function defined in Ω and u(t,z)∈C1(Ω)∩C(Ω). Let p0=(t0,z0) be a fixed point in Ω and Bε(p0)={|p−p0|<ε} be an open ball whose center is p0p_0, and the radius ε is so small that $\overline{B_\varepsilon(p_0)}\subset\Omega$.. Write $\Omega_\varepsilon=\Omega\setminus\overline{B_\varepsilon(p_0)}$. Using the formula (3) in Ωε and replacing *U*, *V* by E(p−p0), u(p) respectively, we have

$$\int_{\Omega_\varepsilon}E(p-p_0)Du\,d\sigma=\int_S E(p-p_0)\Im u(p)\,dS-\int_{\partial B_\varepsilon(p_0)}E(p-p_0)\Im u(p)\,dS. \tag{4}$$

Where

$$\int_{\partial B_\varepsilon(p_0)}E(p-p_0)\Im u(p)\,dS=I_1+I_2+I_3,$$

$$I_1=\frac{\lambda e^{-\lambda\varepsilon}}{4\pi\varepsilon^2}\int_{\partial B_\varepsilon(p_0)}\left[(t-t_0)e_1+(x-x_0)e_2+(y-y_0)e_3\right]u(p)\,dS,$$

$$I_2=-\frac{\lambda e^{-\lambda\varepsilon}}{4\pi\varepsilon}\int_{\partial B_\varepsilon(p_0)}u(p)\,dS,$$

$$I_3=-\frac{e^{-\lambda\varepsilon}}{4\pi\varepsilon^2}\int_{\partial B_\varepsilon(p_0)}u(p)\,dS.$$

It is easy to show that

$$\lim_{\varepsilon\to 0} I_1 = 0, \qquad \lim_{\varepsilon\to 0} I_2 = 0, \qquad \lim_{\varepsilon\to 0} I_3 = -u(p_0).$$

Then letting ε tend to zero in (4), we obtain the following Pompeiu formula corresponding to the operator *D*.

Theorem 1 *Let* Ω *be a bounded domain in* R×C *with piecewise smooth boundary S. If* u(t,z)=u(p) *is a complex vector function defined in* Ω *and* $u(t,z) \in C^1(\Omega)\cap C(\overline{\Omega})$, *then*

$$u(t,z) = -\int_S E(p'-p)\Im u(p')\,dS_{p'} + \int_\Omega E(p'-p)D_{p'}u\,d\sigma_{p'}, \quad (t,z)\in\Omega. \tag{5}$$

By applying Theorem 1, we can deduce the following Cauchy integral formula of the Hλ-regular vector function.

Theorem 2 *If a complex vector function* Ψ(t,z)∈C1(Ω)∩C(Ω¯) *and satisfies the equation* DΨ=0 *in* Ω, *then*

$$\Psi(t,z) = -\int_S E(p'-p)\Im u(p')\,dS_{p'}, \quad (t,z)\in\Omega, \tag{6}$$

and if (t,z)∈¯Ω, *then*

$$\int_S E(p'-p)\Im u(p')\,dS_{p'} = 0. \tag{7}$$

Proof The formula (6) follows directly from the Pompeiu formula (5) and the equality (7) can easily be derived from (3). □

3 CAUCHY TYPE INTEGRAL AND PLEMELJ FORMULA

Let φ(t,z) be a complex vector function defined on a closed smooth surface *S* in R×C, φ(t,z)∈Cα(S), 0<α<1. Denote

$$\Psi(t,z) = -\int_S E(p'-p)\Im\varphi(p')\,dS_{p'}, \tag{8}$$

and call Φ(t,z) the Cauchy type integral with respect to the operator *D*. In the following, we shall simply call it the Cauchy type integral. In addition, φ(t,z) is called the density function of Ψ(t,z).

For arbitrary p=(t,z)∈¯S, there exists a neighborhood Bρ(p) of *p* which does not intersect with *S*. In Bρ(p),

$$\begin{aligned} D\Psi(t,z) &= -(\lambda+\nabla_p)\int_S(\lambda+\nabla_{p'})\left(\frac{1}{4\pi|p'-p|}e^{-\lambda|p'-p|}\right)\Im\varphi(p')\,dS_{p'} \\ &= -(\lambda+\nabla_p)\int_S(\lambda-\nabla_p)\left(\frac{1}{4\pi|p'-p|}e^{-\lambda|p'-p|}\right)\Im\varphi(p')\,dS_{p'} \\ &= -\int_S(\lambda^2-\Delta_p)\left(\frac{1}{4\pi|p'-p|}e^{-\lambda|p'-p|}\right)\Im\varphi(p')\,dS_{p'} = 0. \end{aligned}$$

Consequently, Ψ(t,z) is Hλ-regular in the exterior of *S*. In addition, it is easy to see that Ψ(t,z) converges to 0 as (t,z)→∞.

When (t,z)∈S, we provide that the integral on the right-hand side of (8) represents Cauchy's principal value.

Lemma 1 *Let Ω be a bounded domain in* R×C *with smooth boundary S. If* p∈S$p\in S$*, in the sense of Cauchy's principal value, then*

$$\int_S E(p'-p)\Im\,dS_{p'} = -\frac{1}{2}e_0 + \lambda\int_\Omega E(p'-p)\,d\sigma_{p'}. \tag{9}$$

Proof Let Bε(p) be an open ball with the radius ε and the center *p*, write the component of ∂Bε(p) lying in the exterior of Ω as Γ. Then *x* is an interior point of the region inclosed by the closed surface S′=(S\(S∩Bε(p)))∪Γ. By the Pompeiu formula (5), we have

$$e_0 = -\left(\int_{S\setminus(S\cap B_\varepsilon(p))} + \int_\Gamma\right) E(p'-p)\Im\, dS_{p'} + \lambda \int_{\Omega\cup B_\varepsilon(p)} E(p'-p)\, d\sigma_{p'}. \tag{10}$$

Similarly to the proof of Theorem 1, we can derive

$$\lim_{\varepsilon\to 0}\int_\Gamma E(p'-p)\Im\, dS_{p'} = -\frac{1}{2}e_0.$$

Letting ε→0 in (10), it follows that (9) holds. □

By using Lemma 1, we can obtain the following Plemelj formula of the Cauchy type integral (8).

Theorem 3 *Write the domain* Ω *as* Ω+ *and the complementary domain of* Ω⁻ *as* Ω−. *When* p *tends to* p0(∈S) *from* Ω+ *and* Ω−*respectively, the limits of the Cauchy type integral* (8) *exist, which will be written as* Ψ+(p0) *and* Ψ−(p0) *respectively, and*

$$\begin{cases} \Psi^+(p_0) = -\int_S E(p-p_0)\Im\varphi(p)\, dS_p + \frac{1}{2}\varphi(p_0), \\ \Psi^-(p_0) = -\int_S E(p-p_0)\Im\varphi(p)\, dS_p - \frac{1}{2}\varphi(p_0). \end{cases} \tag{11}$$

The above formula can be rewritten as

$$\begin{cases} \Psi^+(p_0) - \Psi^-(p_0) = \varphi(p_0), \\ \Psi^+(p_0) + \Psi^-(p_0) = -2\int_S E(p-p_0)\Im\varphi(p)\, dS_p. \end{cases} \tag{12}$$

Proof Since $\varphi(p) \in C_\alpha(S)$, 0<α<10 $< \alpha < 1$, therefore the improper integral $\int_S E(p-p_0)\Im(\varphi(p)-\varphi(p_0))\,dS_p$ is convergent. By Lemma 1, we have

$$\int_S E(p-p_0)\Im\varphi(p)\,dS_p = \int_S E(p-p_0)\Im(\varphi(p)-\varphi(p_0))\,dS_p$$
$$-\frac{1}{2}\varphi(p_0) + \lambda\int_\Omega E(p-p_0)\,d\sigma_y \cdot \varphi(x_0).$$

The Cauchy type integral (8) can be written in the following form:

$$-\int_S E(p'-p)\Im\varphi(p')\,dS_{p'} = -\int_S E(p'-p)\Im(\varphi(p')-\varphi(p_0))\,dS_{p'}$$
$$-\int_S E(p'-p)\Im\,dS_{p'} \cdot \varphi(p_0). \tag{13}$$

By the Pompeiu formula, we obtain

$$\int_S E(p'-p)\Im\,dS_{p'} = \begin{cases} -e_0 + \lambda\int_\Omega E(p'-p)\,d\sigma_{p'}, & p \in \Omega^+, \\ \lambda\int_\Omega E(p'-p)\,d\sigma_{p'}, & p \in \Omega^-. \end{cases}$$

When p(∈⁻S)→p0(∈S), using the method similar to one complex variable [12, 13], we can show that

$$\lim_{p\to p_0}\int_S E(p'-p)\Im(\varphi(p')-\varphi(p_0))\,dS_{p'} = \int_S E(p'-p_0)\Im(\varphi(p')-\varphi(p_0))\,dS_{p'}.$$

Moreover, by using the Hölder inequality, it is easy to show that

$$\lim_{p\to p_0}\int_\Omega E(p'-p)\,d\sigma_{p'} = \int_\Omega E(p'-p_0)\,d\sigma_{p'}.$$

Thus letting p tend to p0(∈S)$p_0(\in S)$ from Ω+Ω^+ and Ω−Ω^- respectively in (13), we get

$$\Psi^+(p_0) = -\int_S E(p'-p_0)\Im\big(\varphi(p') - \varphi(p_0)\big)\, dS_{p'} + \varphi(p_0) - \lambda\int_\Omega E(p'-p_0)\, d\sigma_{p'}\cdot\varphi(p_0)$$
$$= -\int_S E(p'-p_0)\Im\varphi(p')\, dS_{p'} + \frac{1}{2}\varphi(p_0),$$

$$\Psi^-(p_0) = -\int_S E(p'-p_0)\Im\big(\varphi(p') - \varphi(p_0)\big)\, dS_{p'} - \lambda\int_\Omega E(p'-p_0)\, d\sigma_{p'}\cdot\varphi(p_0)$$
$$= -\int_S E(p'-p_0)\Im\varphi(p')\, dS_{p'} - \frac{1}{2}\varphi(p_0).$$

This is (11), and (12) is easily deduced from (11). □

The following result follows directly from Theorem 3.

Corollary 1 *Let* Ω *be a bounded domain in* R×C *whose boundary is a closed smooth surface S.* φ(t,z) *is a complex vector function defined on the surface S, and* φ(t,z)∈Cα(S), 0<α<1. *Then the Cauchy type integral* (8) *whose density function is* φ(t,z) *is a Cauchy integral if and only if* (t0,z0)∈S,

$$\Psi^-(t_0, z_0) = 0.$$

4 OPERATOR TΩF

Let f(t,z) be a complex vector function defined in a bounded domain Ω of R×C. Denote

$$T_\Omega f = \int_\Omega E(p'-p) f(p')\, d\sigma_{p'}, \tag{14}$$

where

$$E(p'-p) = \frac{1}{4\pi}\left[\frac{\lambda}{|p'-p|} - \left(\frac{\lambda}{|p'-p|^2} + \frac{1}{|p'-p|^3}\right)\big((t'-t)e_1 + (x'-x)e_2 + (y'-y)e_3\big)\right]e^{-\lambda|p'-p|}.$$

In this section, we shall get that if $f(t,z) \in L_1(\overline{\Omega})$, then TΩf is a distribution solution of the inhomogeneous equation

$$Du = f, \tag{15}$$

and shall discuss some properties of the operator TΩf.

Similarly to the quaternion calculus [3, 17], we can obtain the following results.

Theorem 4 *If* f(t,z) $\in L_1(\overline{\Omega})$, *then* TΩf *exists for all* (t,z) *in the exterior of* Ω. *Beside* TΩf *is* Hλ-*regular in the exterior of* Ω*and*

$$T_\Omega f = 0, \quad (t,z) \to \infty.$$

Theorem 5 *Let* f(t,z)∈L1(Ω¯), *then* TΩf *exists almost everywhere on* R×C$\mathbb{R} \times \mathbb{C}$ *and belongs to* Lp(Ω¯¯¯∗), 1≤p<32, *where* Ω∗Ω∗*denotes any bounded domain in* R×C.

For complex vector functions $f(t,z) = \binom{f_1(t,z)}{f_2(t,z)}, \varphi(x) = \binom{\varphi_1(t,z)}{\varphi_2(t,z)}$ given on Ω, define

$$(f,\varphi) = \int_\Omega (\overline{f_1}\varphi_1 + \overline{f_2}\varphi_2)\, d\sigma.$$

When f(t,z)∈L1(Ω¯), φ(t,z)∈C∞0(Ω), it is easy to show that f(φ)=(f,φ) is a distribution on C∞0(Ω).

Theorem 6 *Let* f(t,z)∈L1(Ω¯). *Then for any* $\varphi(t,z) = \binom{\varphi_1(t,z)}{\varphi_2(t,z)} \in C_0^\infty(\Omega)$,

(TΩf,D′φ)=(f,φ)

holds.

Proof From the equality (2), we get

$$\int_{\Omega}\left[U(D'V)-(UD)V\right]d\sigma=-\int_{S}U\Im V\,dS.$$

In the above equality replacing *U*, *V* by $E'(p'-p)=D'(\frac{1}{4\pi|p'-p|}e^{-\lambda|p'-p|})$, $u(p)$ respectively, by using the method analogous to the proof of Pompeiu formula (5), we can derive the Pompeiu formula corresponding to the operator D', *i.e.*, if $u(t,z)\in C^{1}(\Omega)\cap C(\overline{\Omega})$, then

$$u(t,z)=\int_{S}E'(p'-p)\Im u(p')\,dS_{p'}+\int_{\Omega}E'(p'-p)D'_{p'}u\,d\sigma_{p'},\quad (t,z)\in\Omega. \tag{16}$$

Thus for any $\varphi(t,z)\in C_{0}^{\infty}(\Omega)$,

$$\varphi(t,z)=T'_{\Omega}D'\varphi=\int_{\Omega}E'(p'-p)D'_{p'}\varphi\,d\sigma_{p'}$$

holds.

In accordance with Theorem 5, $T_{\Omega}f\in L_{1}(\overline{\Omega})$. Thereby by the Fubini theorem,

$$\left(T_{\Omega}f,D'\varphi\right)=\left(f,T'_{\Omega}\left(D'\varphi\right)\right)=(f,\varphi),$$

the desired result follows. □

Let complex vector functions f,g∈L1(Ω¯). If for any φ(t,z)∈C∞0(Ω),

$$\left(g,D'\varphi\right)=(f,\varphi),$$

then f is called a generalized derivative corresponding to the operator D of g. The derivative is denoted by f=(g)D. From Theorem 6 and the definition, (TΩf)D=f.

Theorem 7 *If a complex vector function* g∈C¹(Ω)$g\in C^{1}(\Omega)$ *and satisfies the equation* Dg=f, *then*

$(g)_D = f$.

This shows that if the complex vector function g is a classical solution of the equation (15), then it is also a distributional solution of the equation.

Proof It follows by the definition and the divergence theorem. □

Let a=(a1,a2) be a complex vector. The model of *a* is defined

$$|a| = \left(|a_1|^2 + |a_2|^2\right)^{\frac{1}{2}}.$$

It is easy to show that

$$\left|a(te_1 + xe_2 + ye_3)\right| = \left|(a_1, a_2)\begin{pmatrix} t & z \\ \overline{z} & -t \end{pmatrix}\right| = \left|(a_1 t + a_2\overline{z}, a_1 z - a_2 t)\right|$$

$$= \left[\left(|a_1|^2 + |a_2|^2\right)\left(t^2 + |z|^2\right)\right]^{\frac{1}{2}} = |a||te_1 + xe_2 + ye_3|.$$

By using similar methods to those used when proving the Hölder continuous of the operator *T* in quaternion calculus [16, 17], we can prove the following theorem.

Theorem 8 *Let* Ω *be a bounded domain in* R×C, *the complex vector function* f(t,z)∈Lp(Ω⁻), p>3.

(a)For any $\zeta \in \mathbb{R} \times \mathbb{C}$,

$$\left|T_\Omega f(\zeta)\right| \le M(p, \Omega)\|f\|_{L_p}, \tag{17}$$

where M(p,Ω)$M(p,\Omega)$ *is a positive real constant depending only on p,* Ω.

*(b)*g(ζ)=TΩf(ζ) *satisfies*

$$\left|g(\zeta_1) - g(\zeta_2)\right| \le M'(p)\|f\|_{L_p}|\zeta_1 - \zeta_2|^{\alpha}, \qquad \alpha = \frac{p-3}{p}, \zeta_1, \zeta_2 \in \mathbb{R} \times \mathbb{C}, \tag{18}$$

where M(p) *is a positive real constant depending only on p.*

The inequalities (17) *and* (18) *imply that* TΩf *is a compact mapping from* Lp(Ω¯), p>3 *into* $C_\alpha(\overline{\Omega}),\ \alpha = \frac{p-3}{p}$, *and*

$$\|T_\Omega f\|_{C_\alpha(\overline{\Omega})} \leq M\|f\|_{L_p}, \quad p > 3, \alpha = \frac{p-3}{p}. \tag{19}$$

Proof (a) From the definition (14) of TΩf,

$$|T_\Omega f| \leq \frac{\lambda}{2\pi}\int_\Omega \frac{e^{-\lambda|\zeta'-\zeta|}}{|\zeta'-\zeta|}|f(\zeta')|\,d\sigma_{\zeta'} + \frac{1}{4\pi}\int_\Omega \frac{e^{-\lambda|\zeta'-\zeta|}}{|\zeta'-\zeta|^2}|f(\zeta')|\,d\sigma_{\zeta'} = I_1 + I_2.$$

Since

$$e^{-\lambda|\zeta'-\zeta|} \leq \frac{1}{\lambda|\zeta'-\zeta|},$$

we have by Hölder's inequality

$$I_1 \leq \frac{1}{2\pi}\left(\int_\Omega \frac{1}{|\zeta'-\zeta|^{2q}}\,d\sigma_{\zeta'}\right)^{\frac{1}{q}}\|f\|_{L_p},$$

where $\frac{1}{p}+\frac{1}{q}=1$. By hypothesis p>3, we have $q<\frac{3}{2}$. Let d=sup $_{\zeta,\zeta'\in\overline{\Omega}}|\zeta-\zeta'|,\ d_1 = \mathrm{dist}(\zeta,\overline{\Omega})$, namely *d*, d1 denote the diameter of a bounded domain Ω and the distance between ζ and Ω¯ respectively.

If $\zeta \in \overline{\Omega}$,

$$\left(\int_\Omega \frac{1}{|\zeta'-\zeta|^{2q}}\,d\sigma_{\zeta'}\right)^{\frac{1}{q}} \leq \left(\int_{|\zeta'-\zeta|\leq d} \frac{1}{|\zeta'-\zeta|^{2q}}\,d\sigma_{\zeta'}\right)^{\frac{1}{q}} = \left(\frac{4\pi d^{3-2q}}{3-2q}\right)^{\frac{1}{q}}.$$

If $\zeta \overline{\in} \overline{\Omega}$,

$$\left(\int_{\Omega}\frac{1}{|\zeta'-\zeta|^{2q}}d\sigma_{\zeta'}\right)^{\frac{1}{q}}\le\left(\int_{d_1\le|\zeta'-\zeta|\le d_1+d}\frac{1}{|\zeta'-\zeta|^{2q}}d\sigma_{\zeta'}\right)^{\frac{1}{q}}$$
$$=\left(\frac{4\pi d^{3-2q}}{3-2q}\right)^{\frac{1}{q}}\left[\left(1+\frac{d_1}{d}\right)^{3-2q}-\left(\frac{d_1}{d}\right)^{3-2q}\right]^{\frac{1}{q}}$$
$$\le\left(\frac{4\pi d^{3-2q}}{3-2q}\right)^{\frac{1}{q}}.$$

The last inequality is immediate from

$$0<\left(1+\frac{d_1}{d}\right)^{3-2q}-\left(\frac{d_1}{d}\right)^{3-2q}\le 1.$$

In fact, from 0<β=3−2q<1, it is easy to see that the real function μ(x)=(1+x)β−xβ is a monotone decreasing function in [0,+∞) and μ(0)=0, so that 0<μ(x)≤1.

Let

$$M_1(p,\Omega)=\frac{1}{2\pi}\left(\frac{4\pi d^{3-2q}}{3-2q}\right)^{\frac{1}{q}}.$$

Hence we obtain

$$I_1\le M_1(p,\Omega)\|f\|_{L_p}. \tag{20}$$

Noting

$$e^{-\lambda|\zeta'-\zeta|}\le 1,$$

thus

$$I_2 = \frac{1}{4\pi}\int_\Omega \frac{e^{-\lambda|\zeta'-\zeta|}}{|\zeta'-\zeta|^2}|f(\zeta')|\,d\sigma_{\zeta'}$$
$$\le \frac{1}{4\pi}\left(\int_\Omega \frac{1}{|\zeta'-\zeta|^{2q}}\,d\sigma_{\zeta'}\right)^{\frac{1}{q}}\|f\|_{L_p} \le \frac{1}{4\pi}\left(\frac{4\pi d^{3-2q}}{3-2q}\right)^{\frac{1}{q}}\|f\|_{L_p},$$

i.e.,

$$I_2 \le \frac{1}{2}M_1(p,\Omega)\|f\|_{L_p}. \tag{21}$$

The inequality (17) follows immediately from (20) and (21).

*(b)*Without loss of generality, we may take $|\zeta 1-\zeta 2|<1$. We write

$$g(\zeta) = \frac{\lambda}{4\pi}\int_\Omega \frac{e^{-\lambda|\zeta'-\zeta|}}{|\zeta'-\zeta|}f(\zeta')\,d\sigma_{\zeta'} - \frac{\lambda}{4\pi}\int_\Omega \frac{(\zeta'-\zeta)e^{-\lambda|\zeta'-\zeta|}}{|\zeta'-\zeta|^2}f(\zeta')\,d\sigma_{\zeta'}$$
$$-\frac{1}{4\pi}\int_\Omega \frac{(\zeta'-\zeta)e^{-\lambda|\zeta'-\zeta|}}{|\zeta'-\zeta|^3}f(\zeta')\,d\sigma_{\zeta'}$$
$$= g_1(\zeta) - g_2(\zeta) - g_3(\zeta),$$

where

$$\zeta'-\zeta = (t'-t)e_1 + (x'-x)e_2 + (y'-y)e_3.$$

It is easy to see that $(\zeta'-\zeta)2=|\zeta'-\zeta|2$.

We have

$$|g_1(\zeta_1) - g_1(\zeta_2)|$$
$$= \frac{\lambda}{4\pi}\left|\int_\Omega \frac{(|\zeta'-\zeta_2|-|\zeta'-\zeta_1|)e^{-\lambda|\zeta'-\zeta_1|} + |\zeta'-\zeta_1|(e^{-\lambda|\zeta'-\zeta_1|}-e^{-\lambda|\zeta'-\zeta_2|})}{|\zeta'-\zeta_1||\zeta'-\zeta_2|}f(\zeta')\,d\sigma_{\zeta'}\right|$$
$$\le \frac{\lambda}{4\pi}\int_\Omega \frac{|\zeta_1-\zeta_2|(1+\lambda|\zeta'-\zeta_1|)}{|\zeta'-\zeta_1||\zeta'-\zeta_2|}|f(\zeta')|\,d\sigma_{\zeta'}.$$

Here we use the estimates

$$e^{-\lambda|\zeta'-\zeta_1|} \leq 1$$

and

$$\left|e^{-\lambda|\zeta'-\zeta_1|} - e^{-\lambda|\zeta'-\zeta_2|}\right| \leq \lambda\left||\zeta'-\zeta_2| - |\zeta'-\zeta_1|\right| \leq \lambda|\zeta_1-\zeta_2|.$$

We get by Hölder's inequality

$$\int_\Omega \frac{1}{|\zeta'-\zeta_1||\zeta'-\zeta_2|}|f(\zeta')|\,d\sigma_{\zeta'} \leq \left(\int_\Omega \frac{1}{|\zeta'-\zeta_1|^q|\zeta'-\zeta_2|^q}\,d\sigma_{\zeta'}\right)^{\frac{1}{q}} \|f\|_{L_p},$$

and

$$\int_\Omega \frac{1}{|\zeta'-\zeta_2|}|f(\zeta')|\,d\sigma_{\zeta'} \leq \left(\int_\Omega \frac{1}{|\zeta'-\zeta_2|^q}\,d\sigma_{\zeta'}\right)^{\frac{1}{q}} \|f\|_{L_p}.$$

Using the inequality

$$\int_\Omega \frac{1}{|\zeta'-\zeta_1|^\alpha|\zeta'-\zeta_2|^\beta}\,d\sigma_{\zeta'} \leq \begin{cases} c_1|\zeta_1-\zeta_2|^{3-\alpha-\beta}, & 0<\alpha,\beta<3, \alpha+\beta>3, \\ c_2, & 0\leq\alpha,\beta<3, \alpha+\beta<3, \end{cases}$$

where c1c_1, c2c_2 are positive real constants, and noting $q < \frac{3}{2}$, we then obtain

$$\left|g_1(\zeta_1)-g_1(\zeta_2)\right| \leq M_1'(p)\|f\|_{L_p}|\zeta_1-\zeta_2|. \tag{22}$$

By simple computation we have

$$\left|g_2(\zeta_1)-g_2(\zeta_2)\right| = \frac{\lambda}{4\pi}\left|\int_\Omega \frac{(\zeta_1-\zeta_2)e^{-\lambda|\zeta'-\zeta_1|} + (\zeta'-\zeta_1)(e^{-\lambda|\zeta'-\zeta_1|} - e^{-\lambda|\zeta'-\zeta_2|})}{(\zeta'-\zeta_1)(\zeta'-\zeta_2)} f(\zeta')\,d\sigma_{\zeta'}\right|,$$

and

$$\left|g_3(\zeta_1)-g_3(\zeta_2)\right| = \frac{1}{4\pi}\left|\int_\Omega \left\{ \frac{[|\zeta'-\zeta_2|(\zeta_1-\zeta_2)+(\zeta'-\zeta_1)(|\zeta'-\zeta_2|-|\zeta'-\zeta_1|)]e^{-\lambda|\zeta'-\zeta_1|}}{|\zeta'-\zeta_1||\zeta'-\zeta_2|(\zeta'-\zeta_1)(\zeta'-\zeta_2)} + \frac{(\zeta'-\zeta_1)(e^{-\lambda|\zeta'-\zeta_1|}-e^{-\lambda|\zeta'-\zeta_2|})}{|\zeta'-\zeta_2|(\zeta'-\zeta_1)(\zeta'-\zeta_2)} \right\} f(\zeta')\,d\sigma_{\zeta'}\right|.$$

By using a similar method, we can obtain

$$\left|g_2(\zeta_1)-g_2(\zeta_2)\right| \le M_2'(p)\|f\|_{L_p}|\zeta_1-\zeta_2| \tag{23}$$

and

$$\left|g_3(\zeta_1)-g_3(\zeta_2)\right| \le M_3'(p)\|f\|_{L_p}|\zeta_1-\zeta_2|^{\alpha}. \tag{24}$$

The required estimate then follows by combining the resulting inequalities. □

5 SOME BOUNDARY VALUE PROBLEMS FOR HΛH_λ-REGULAR FUNCTIONS

It is well known that the Dirichlet problem for analytic functions w(z)$w(z)$ in a bounded domain of the complex plane, boundary value of which is a given complex value function, is overdetermined, thereby being unsolvable in general. In the theory of boundary value problems for analytic functions, the boundary condition is replaced by Rew=r(t)$\mathrm{Re}\,w = r(t)$, and a more general problem is the so-called Riemann-Hilbert problem with boundary condition Reλ(t)¯w=r(t). Analogously to this, the Dirichlet problem for HλH_λ-regular functions, boundary value of which is a given complex value vector function, is also overdetermined, and we have therefore to consider new boundary conditions. In this section, we introduce and discuss some Riemann-Hilbert type boundary value problems for HλH_λ-regular vector functions.

Let Ω be a bounded domain with smooth boundary *S* in R×C 0∈Ω. *S* satisfies the exterior sphere condition, that is, for every point ζ∈S, there exists a ball *B* satisfying B¯∩Ω¯ζ. Ω0 denotes the transversal domain of Ω on the plane t=0$t=0$, its boundary L=∂Ω0 is a closed smooth curve and the projection of every point of Ω on the plane t=0$t=0$ is in Ω0Ω_0. We consider the following boundary value problems:

Find a continuous solution $u(t,z)=\binom{u_1(t,z)}{u_2(t,z)}$ of the equation

$$Du=f \tag{25}$$

in Ω¯, satisfying the boundary conditions

$$u_1|_S=\varphi, \tag{26}$$

$$u_1|_S=\varphi,$$
$$\operatorname{Re}\overline{\lambda(\tau)}u_2(0,\tau)=r(\tau),\quad \tau\in L, \tag{27}$$

where φ is a given complex value function on *S*, φ∈C(S), λ(τ) is a given complex value function on *L*, λ(τ)≠0, τ∈L. r(τ) is a given real value function on *L*, λ(τ), r(τ)∈Cα, 0<α<1. This problem is called problem H of the equation (25), and $\kappa=\frac{1}{2\pi}\Delta_L\arg\lambda(\tau)$ is called index of the problem H.

When κ=0$\kappa=0$, if *u* satisfies the condition

$$\operatorname{Im}u_2(0,0)=a \tag{28}$$

besides the above boundary conditions, where *a* is a real constant, then the problem is called problem D.

In particular, when f=0 in the equation (25), the above problems are namely the problem H and problem D for the Hλ-regular vector functions.

Lemma 2 *Suppose complex value functions*

$g_1(t,z),\ g_2(t,z) \in C(\overline{\Omega}) \cap C^1(\Omega)$. *If* $g_1(t,z)$, g2(t,z) *satisfy compatible condition*

$$\left(\lambda - \frac{\partial}{\partial t}\right) g_1 = \frac{\partial}{\partial \overline{z}} g_2, \tag{29}$$

then the following overdetermined system with respect to u(t,z)

$$\begin{cases} \dfrac{\partial}{\partial \overline{z}} u = g_1(t,z), \\ \left(\lambda - \dfrac{\partial}{\partial t}\right) u = g_2(t,z) \end{cases} \tag{30}$$

has the general solution

$$u(t,z) = T_{\Omega_0} g_1 + \Phi(z) e^{\lambda t} - F(t,z) e^{\lambda t}, \tag{31}$$

here Φ(z) *is any analytic function in* Ω0,

$$T_{\Omega_0} g_1 = -\frac{1}{\pi} \int_{\Omega_0} \frac{g_1(t,\zeta)}{\zeta - z} d\sigma_\zeta,$$

$$F(t,z) = \int_0^t e^{-\lambda \xi} \left(\frac{1}{2\pi i} \int_\Gamma \frac{g_2(\xi,\zeta)}{\zeta - z} d\zeta \right) d\xi.$$

Proof Noting the compatible condition and that F(t,z)$F(t,z)$ is an analytic function with respect to z, using the Pompeiu formula [12], it is not difficult to verify by direct calculation that u(t,z)$u(t,z)$ expressed by (31) is the general solution of the system (30). □

As a special case of Theorem 6.13 in [4], we can derive the following result.

Lemma 3 *If φ is continuous on S, then the Dirichlet problem with the boundary condition*

$$w|_S = \varphi$$

for the equation $(\lambda^2-\Delta)w=0$ *in* Ω *has a unique solution* $w(t,z)\in C(\bar{\Omega})\cap C^{2}_{\alpha}(\Omega)$.

Similarly to harmonic function, we have the following result.

Lemma 4 *For the Dirichlet problem of the equation* $(\lambda^2-\Delta)w=0$ *in* Ω, *Green functions* $G(p,p')$ *exist such that the solutions of the problem can be represented by* $G(p,p')$, *namely we have*

$$w(p) = -\int_S w(p')\frac{\partial G}{\partial \nu}\, dS_{p'}. \tag{32}$$

These Green functions $G(p,p')$ *are unique.*

Proof Suppose functions $w,v\in C^1(\bar{\Omega})\cap C^2(\Omega)$. By Green's second identity

$$\int_\Omega (w\Delta v - v\Delta w)\, d\sigma = \int_S \left(w\frac{\partial v}{\partial \nu} - v\frac{\partial w}{\partial \nu}\right) dS,$$

where ν denotes the unit outward normal to the surface S, we obtain

$$\int_\Omega \{v[(\lambda^2-\Delta)w - w(\lambda^2-\Delta)v]\}\, d\sigma = \int_S \left(w\frac{\partial v}{\partial \nu} - v\frac{\partial w}{\partial \nu}\right) dS. \tag{33}$$

Let p be a fixed point in Ω and $B_\varepsilon(p)=\{|p'-p|<\varepsilon\}$ be an open ball whose radius ε is so small that $\overline{B_\varepsilon(p)} \subset \Omega$. Write $\Omega_\varepsilon = \Omega \setminus \overline{B_\varepsilon(p)}$.

Replacing v by $\frac{1}{4\pi}\frac{e^{-\lambda|p'-p|}}{|p'-p|}$, using the formula (33) in Ωε and letting ε tend to zero, similarly to the proof of Theorem 1, we can derive

$$w(p) = -\frac{1}{4\pi}\int_S\left[w\frac{\partial}{\partial\nu}\left(\frac{e^{-\lambda|p'-p|}}{|p'-p|}\right) - \left(\frac{e^{-\lambda|p'-p|}}{|p'-p|}\right)\frac{\partial w}{\partial\nu}\right]dS + \frac{1}{4\pi}\int_\Omega \frac{e^{-\lambda|p'-p|}}{|p'-p|}(\lambda^2-\Delta)w\,d\sigma.$$

Thus when w∈C1(Ω¯)∩C2(Ω) and satisfies the equation (λ2−Δ)w=0,

$$w(p) = -\frac{1}{4\pi}\int_S\left[w\frac{\partial}{\partial\nu}\left(\frac{e^{-\lambda|p'-p|}}{|p'-p|}\right) - \left(\frac{e^{-\lambda|p'-p|}}{|p'-p|}\right)\frac{\partial w}{\partial\nu}\right]dS. \tag{34}$$

For a given p in Ω, find g(p′,p) which satisfies the equation (λ2−Δ)g=0 in Ω and the boundary condition $g(p',p) = \frac{1}{4\pi}\frac{e^{-\lambda|p'-p|}}{|p'-p|}$ on *S*. By virtue of Lemma 3, this g(p′,p) is existential and unique. Write $G(p,p') = \frac{1}{4\pi}\frac{e^{-\lambda|p'-p|}}{|p'-p|} - g(p',p).$ When w satisfies the equation (λ2−Δ)w=0 in Ω, from (33) we derive

$$\int_S\left(w\frac{\partial g}{\partial\nu} - g\frac{\partial w}{\partial\nu}\right)dS = 0.$$

Subtracting this from (34), we get

$$w(p) = -\int_S w(p')\frac{\partial G}{\partial\nu}\,dS_{p'}.$$

A simple approximation argument shows that this formula continues to hold for w∈ $C(\overline{\Omega})\cap C^2(\Omega)$.□

With the aid of the methods of conformal mapping and standardizing boundary condition from complex analysis (see [12, 13]), we can map conformally Ω0Ω_0 into the unit disk on the plane t=0$t=0$, and transform λ(τ)$\lambda(\tau)$ in the condition (33) into τκτ^κ. Hence without loss of generality, we shall directly suppose that Ω0 is the unit disk B1(0) on the plane t=0 and replace (27) by the following condition

$$\mathrm{Re}\,\overline{\tau}^{\kappa} u_2(0,\tau) = r(\tau), \quad \tau \in L = \partial B_1(0). \tag{35}$$

Using these results, we can discuss the solvability of the problem H and the problem D for the HλH_λ-regular vector functions and the equation Du=f.

Theorem 9 (1) *If the index* κ≥0, *the problem H for the* HλH_λ*-regular vector functions in* Ω *is solvable. The problem has the general solutions* $\Psi(t,z) = \begin{pmatrix} \psi_1(t,z) \\ \psi_2(t,z) \end{pmatrix}$, *with*

$$\psi_1(t,z) = -\int_S \varphi(p')\frac{\partial G}{\partial \nu}\, dS_{p'}, \tag{36}$$

$$\psi_2(t,z) = T_{B_1(0)}g_1 + \Phi(z)e^{\lambda t} - F(t,z)e^{\lambda t}, \tag{37}$$

where

$$g_1 = \frac{1}{2}\left(\lambda + \frac{\partial}{\partial t}\right)\psi_1,$$

$$g_2(t,z) = -2\frac{\partial}{\partial z}\psi_1,$$

$$\Phi(z) = -\frac{1}{\pi}\int_{B_1(0)} \frac{z^{2\kappa+1}\overline{g_1(0,\zeta)}}{1-\overline{\zeta}z}\, d\sigma_\zeta + \frac{z^\kappa}{2\pi i}\int_L r(\tau)\frac{\tau+z}{\tau-z}\frac{d\tau}{\tau} + \sum_{m=0}^{2\kappa} c_m z^m,$$

here cmc_m, m=0,…,κ *are arbitrary complex constants, satisfying*

c2κ−m+cm¯=0.

(2)If the index κ<0$\kappa<0$, *the problem H for the* HλH_λ*-regular vector functions in* Ω *is solvable if and only if the function* r(τ)$r(\tau)$*in the boundary conditions* (27) *satisfies the following conditions*

$$\mathrm{Re}\left[\frac{1}{2\pi i}\int_L \frac{r(\tau)}{\tau}d\tau - \frac{1}{\pi}\int_{B_1(0)} \zeta^{-\kappa-1} g_1(0,\zeta)\,d\sigma_\zeta\right] = 0,$$

$$\frac{1}{\pi i}\int_L \frac{r(\tau)}{\tau^{m+1}}d\tau - \frac{1}{\pi}\int_B [\zeta^{-\kappa-m-1} g_1(0,\zeta) + \overline{\zeta}^{-\kappa+m-1}\overline{g_1(0,\zeta)}]\,d\sigma_\zeta = 0,$$

$$m = 1,\ldots,-\kappa-1. \tag{38}$$

When the conditions (38) *hold, the solution then has the same expression as* (1), *except that*

$$\Phi(z) = -\frac{1}{\pi}\int_{B_1(0)} \frac{\overline{\zeta}^{-2\kappa-1}\overline{g_1(0,\zeta)}}{1-\overline{\zeta}z}d\sigma_\zeta + \frac{1}{\pi i}\int_L \frac{r(\tau)}{\tau^{-\kappa}(\zeta - z)}d\tau.$$

Proof $\Psi(t,z) = \binom{\psi_1(t,z)}{\psi_2(t,z)}$ is a HλH_λ-regular vector function, then Ψ(t,z) satisfies the equation DΨ=0$D\Psi = 0$ which is equivalent to

$$\begin{cases} \dfrac{\partial}{\partial \overline{z}}\psi_2 = -\dfrac{1}{2}\left(\lambda + \dfrac{\partial}{\partial t}\right)\psi_1 = g_1(t,z), \\ \left(\lambda - \dfrac{\partial}{\partial t}\right)\psi_2 = -2\dfrac{\partial}{\partial z}\psi_1 = g_2(t,z). \end{cases} \tag{39}$$

From Lemma 4, the function ψ1(t,z) expressed in (36) is the unique solution of the Dirichlet problem with the boundary condition (26) for the equation (λ2−∆)w=0 in Ω, so that g1(t,z), g2(t,z) satisfy the compatible condition of Lemma 2

$$\left(\lambda - \frac{\partial}{\partial t}\right)g_1 = \frac{\partial}{\partial \overline{z}}g_2. \tag{40}$$

Consequently, ψ2(t,z)$\psi_2(t,z)$ can be given by the formula (37). Furthermore, ψ2(t,z)$\psi_2(t,z)$ expressed in (37) satisfies the boundary condition (35) if and only if the analytic function Φ(z)$\Phi(z)$ satisfies the following boundary condition

$$\operatorname{Re}\overline{\tau}^{\kappa}\Phi(\tau) = r(\tau) - \operatorname{Re}\overline{\tau}^{\kappa}\, T_{B_1(0)}g_1(0,\tau), \quad \tau \in L. \tag{41}$$

By means of the results about the Riemann-Hilbert boundary value problem for analytic function in the unit disk [13], we can derive the solvable conditions and the expression of solutions. □

Corollary 2 *The problem D for the* Hλ*-regular vector functions in* Ω *has a unique solution, and the solution is* $\Psi(t,z) = \binom{\psi_1(t,z)}{\psi_2(t,z)}$ *which* ψ1(t,z) *is given by* (36) *and* ψ2(t,z) *expressed as* (37) *where*

$$\Phi(z) = -\frac{1}{\pi}\int_{B_1(0)} \frac{z\overline{g_1(0,\zeta)}}{1-\overline{\zeta}z}\,d\sigma_\zeta + \frac{1}{2\pi i}\int_L r(\tau)\frac{\tau+z}{\tau-z}\frac{d\tau}{\tau} + i\big(a - \operatorname{Im} T_{B_1(0)}g_1(0,z)\big).$$

Proof The result follows immediately from Theorem 9 and the results of the Dirichlet boundary value problem for analytic function in the unit disk. □

Since the solution u of the equation Du=f can be expressed as u=Ψ+TΩf, where Ψ is any HλH_λ-regular vector functions in Ω, if f∈Lp(Ω¯), p>3, then TΩf∈Cα(Ω¯), therefore the problem H of the equation Du=f$Du=f$ in Ω can be transformed into the problem H of the HλH_λ-regular vector function $\Psi(t,z) = \binom{\psi_1(t,z)}{\psi_2(t,z)}$ in Ω with the following boundary conditions

$$\psi_1(t,z) = \varphi_1(t,z) = \varphi(t,z) - T^1_\Omega f, \quad (t,z) \in S,$$

$$\operatorname{Re}\overline{\tau}^{\kappa}\psi_2(0,\tau) = r_1(\tau) = r(\tau) - \operatorname{Re}\overline{\tau}^{\kappa}\, T^2_\Omega f, \quad \tau \in L,$$

where

$$T^1_\Omega f = \frac{1}{4\pi}\int_\Omega \Bigg\{ \frac{\lambda e^{-\lambda|\zeta'-\zeta|}}{|\zeta'-\zeta|} f_1 - \left(\frac{\lambda}{|\zeta'-\zeta|^2} + \frac{1}{|\zeta'-\zeta|^3}\right) e^{-\lambda|\zeta'-\zeta|}\big[(t'-t)f_1 + (z'-z)f_2\big]\Bigg\}\, d\sigma_{\zeta'},$$

namely $T_\Omega f = \binom{T^1_\Omega f}{T^2_\Omega f}$ Using Theorem 10, we obtain the following result about the problem H for the equation Du=f in Ω.

Theorem 10 *Let* f∈Lp(Ω⁻), p>3.

(*a*) *If the index* κ≥0, *the problem H for the equation* Du=f *in* Ω *has the solution* u(t,z)=Ψ(t,z)+TΩf, *where the* Hλ*-regular vector function* Ψ(t,z) *is expressed as* (*a*) *of Theorem* 9 *with* φ1(t,z), r1(τ) *replacing* φ(t,z), r(τ) *respectively.*

(*b*) *If the index* κ<0, *replacing* φ(t,z) *by* φ1(t,z), *the problem H for the equation* Du=f *in* Ω *is solvable if and only if the function* r1(τ)$r_1(\tau)$ *satisfies the conditions* (38). *When the conditions* (38) *hold, the problem then has the solution* u(t,z)=Ψ(t,z)+TΩf, *where the* HλH_λ*-regular vector function* Ψ(t,z) *is expressed as* (*b*) *of Theorem* 9 *with* φ1(t,z), r1(τ) *replacing* φ(t,z), r(τ) *respectively.*

In the same way, we can obtain the result about the problem D for the equation Du=f in Ω.

Corollary 3 *Suppose that* f∈Lp(Ω⁻),p>3. *The problem D for the equation* Du=f *in* Ω *has a unique solution* u(t,z)=Ψ(t,z)+TΩf, *where the* HλH_λ*-regular vector function* Ψ(t,z) *is expressed as Corollary* 2 *with* φ1(t,z), r1(τ) *and* a1=a−ImT2Ωf *replacing* φ(t,z), r(τ)$r(\tau)$ *and a respectively.*

ACKNOWLEDGEMENT

This work is supported by National Natural Science Foundation of China (61173121), the Foundation of Doctor Education of China (20095134110001), and the Key Project Foundation of the Education Department of Sichuan Province of China (12ZA136). The authors would like to thank the referee for helpful comments and suggestions.

COMPETING INTERESTS

The authors declare that they have no competing interests.

AUTHORS' CONTRIBUTIONS

PWY has presented the main purpose of the article. Both authors read and approved the final version of the manuscript.

REFERENCES

1. Caçåo I, Constales D, Kraußhar RS: A unified approach for the treatment of some higher dimensional Dirac type equations on spheres.*Proceedings of the 18th International Conference on the Application of Computer Science and Mathematics in Architecture and Civil Engineering, 07-09 July* 2009.

2. Cürlebeck K: Hypercomplex factorization of the Helmholtz equation. *Z. Anal. Anwend.* 1986, 5(2):125-131.

3. Cürlebeck K, Sprößig W: *Quaternionic Analysis and Elliptic Boundary Value Problem*. Birkhäuser, Boston; 1990.

4. Gilbarg D, Trudinger NS: *Elliptic Partial Differential Equations of Second Order.* Springer, Berlin; 2001.

5. Ha TVN: Integral representation in quaternionic analysis related to the Helmholtz operator. *Complex Var. Theory Appl.* 2003, 48: 1005-1021.

6. Huang LD: The modified Helmholtz equations and their boundary value problems. *Acta Math. Sci.* 1988, 8(2):165-179.

7. Huang LD: The existence and uniqueness theorems of the linear and nonlinear Riemann-Hilbert problems for the generalized holomorphic vectors of the second kind. *Acta Math. Sci.* 1990, 10(2):185-199.

8. Kravchenko VV: *Applied Quaternionic Analysis*. Heldermann, Berlin; 2003.

9. Kravchenko VV, Sharpiro MV: Helmholtz operator with a quaternionic wave number and associated function theory, II. Integral representations. *Acta Appl. Math.* 1993, 32(3):243-265.

10. Kravchenko VV, Sharpiro MV: Hypercomplex factorization of the multidimensional Helmholtz operator and some it's applications. *Dokl. Sem. Inst. I. N. Vekua, Tbilisi* 1990, 5(1):106-109.

11. Sharpiro, MV: On the properties of a class of singular integral equations connected with the three-dimensional Helmholtz equation. In: Abstracts of lectures at the 14 school on Operator Theory in Functional Spaces, Novgorod, U.S.S.R. (1989)

12. Vekya IN: *Generalized Analytic Functions*. Pergamon, Oxford; 1962.

13. Wen GC: *Conformal Mappings and Boundary Value Problems*. Amer. Math. Soc., Providence; 1992. Translated from the Chinese by Kuniko Weltin

14. Xu ZY:A function theory for the operator $(D-\lambda)(D-\lambda)$. *Complex Var. Theory Appl.* 1991, 16(1):27-42.

15. Xu ZY: Helmholtz equations and boundary value problems. Pitman Res. Notes Math. Ser. 262. In *Partial Differential Equations with Complex Analysis*. Longman Sci. Tech., Harlow; 1992:204-214.

16. Yang PW:Hölder continuity of TGf$T_G f$ and Riemann-Hilbert boundary value problem in quaternion calculus. *Acta Math. Sin. Chinese Ser.*2003, 46(5):993-998.

17. Yang PW: *Quaternion Calculus and Partial Differential Equations.* Science Press, Beijing; 2009.

18. Yang PW, Li ML, Chen Y: Initial-boundary value problems of some hyperbolic systems of first order equations. *Acta Math. Appl. Sin.* 2008, 31(1):61-70.

1. Yang PW, Yang S, Li ML: An initial-boundary value problem for the Maxwell equations. *J. Differ. Equ.* 2010, 249: 3003-3023.

4

ON CHEN INVARIANTS AND INEQUALITIES IN QUATERNIONIC GEOMETRY

Gabriel-Eduard Vîlcu*

Department of Mathematical Modelling, Economic Analysis and Statistics, Petroleum-Gas University of Ploie¸sti, Bulevardul Bucure¸sti Nr. 39, Ploie¸sti, 100680, Romania Faculty of Mathematics and Computer Science, Research Center in Geometry, Topology and Algebra, University of Bucharest, Str. Academiei, Nr. 14, Sector 1, Bucure¸sti, 70109, Romania

ABSTRACT

In this paper we give a survey of main results concerning Chen inequalities for submanifolds in quaternionic space forms and propose some open problems in the field for further research.

KEYWORDS

Chen's invariant squared, mean curvature shape, operator, quaternionic space form slant submanifold

1 INTRODUCTION

The theory of Chen invariants, initiated by Prof. B.-Y. Chen in a seminal paper published in 1993 [1], is presently one of the most interesting research topic in differential geometry of submanifolds. The author's original motivation to introduce new types of Riemannian invariants, known as δ-invariants or Chen invariants, was the need to provide answers to an open question raised by Chern concerning the existence of minimal immersions into a Euclidean space of arbitrary dimension [2]. In fact, due the lack of control of the extrinsic properties of the submanifolds by the known intrinsic invariants, no solutions to Chern's problem were known before the invention of Chen invariants (see [3]). Therefore, in [1], Chen obtained a necessary condition for the existence of minimal isometric immersion from a given Riemannian manifold M into Euclidean space and established a sharp inequality for a submanifold in a real space form using the scalar curvature and the sectional curvature (both being intrinsic invariants) and squared mean curvature (the main extrinsic invariant). On the other hand, in [4], Chen obtained inequalities between the k-Ricci curvature, the squared mean curvature and the shape operator for submanifolds in real space forms with arbitrary codimensions. These inequalities are also sharp, and many nice classes of submanifolds realize equality in all above inequalities. Since then many papers concerning Chen invariants and inequalities have appeared in the literature for different classes of submanifolds in various ambient spaces, like complex space forms [5–7], cosymplectic space forms [8–10], Sasakian space forms [11–13], locally conformal Kähler space forms [14–16], generalized complex space forms [17–19], locally conformal almost cosymplectic manifolds [20, 21], (κ,μ)$^{(\kappa,\mu)}$-contact space forms [22, 23], Kenmotsu space forms [24, 25], S-space forms [26, 27], T-space forms [28]; see also [29] and references therein.

In the last decade, Chen-like inequalities were extended in the quaternionic setting. It is the main purpose of this paper to present the evolution of these inequalities for submanifolds in quaternionic space forms, to survey some recent results on this topic and to propose a set of natural problems in the field.

2 PRELIMINARIES

2.1 Riemannian invariants

In this subsection we recall some basic concepts concerning Chen invariants, using mainly [30].

Let *M* be an *n*-dimensional Riemannian manifold. We denote by K(π) the sectional curvature of *M* associated with a plane section π⊂TpM, p∈M. If {e1,...,en} is an orthonormal basis of the tangent space TpM, the scalar curvature τ at *p* is defined by

$$\tau(p) = \sum_{1\leq i<j\leq n} K(e_i \wedge e_j). \tag{1}$$

One denotes

$$(\inf K)(p) = \inf\{K(\pi) | \pi \subset T_pM, \dim \pi = 2\} \tag{2}$$

and the Chen first invariant is given by

$$\delta_M(p) = \tau(p) - (\inf K)(p). \tag{3}$$

Suppose *L* is an *r*-dimensional subspace of TpM, r≥2 and {e1,...,er} an orthonormal basis of *L*. We define the scalar curvature τ(L) of the *r*-plane section *L* by

$$\tau(L) = \sum_{1\leq \alpha<\beta\leq r} K(e_\alpha \wedge e_\beta). \tag{4}$$

For an integer k≥0, we denote by S(n,k) the set of *k*-tuples (n1,...,nk) of integers ≥2 satisfying n1<n, n1+⋯+nk≤n. We denote by S(n) the set of unordered *k*-tuples with k≥0 for a fixed *n*.

For each *k*-tuples (n1,...,nk)∈S(n), Chen introduced a Riemannian invariant δ(n1,...,nk) defined by

$$\delta(n_1,\ldots,n_k)(p) = \tau(p) - S(n_1,\ldots,n_k)(p), \tag{5}$$

where

$$S(n_1,\ldots,n_k)(p) = \inf\{\tau(L_1) + \cdots + \tau(L_k)\}, \tag{6}$$

L1,...,Lk running over all *k* mutually orthogonal subspaces of TpM such that dimLj=nj, j∈{1,...,k}. We note that the Chen invariant with k=0$k=0$ is nothing but the scalar curvature. Also, we denote by d(n1,...,nk) and b(n1,...,nk) the real constants given by

$$d(n_1,\ldots,n_k) = \frac{n^2(n+k-1-\sum_{j=1}^{k} n_j)}{2(n+k-\sum_{j=1}^{k} n_j)}$$

and

$$b(n_1,\ldots,n_k) = \frac{n(n-1)-\sum_{j=1}^{k} n_j(n_j-1)}{2}.$$

For a *k*-plane section *L* of TpM, p∈M, and *X* a unit vector in *L*, one can choose an orthonormal basis {e1,...,ek} of *L* such that e1=X$e_1 = X$. Then the Ricci curvature of *L* at *X*, denoted RicL(X), is defined by

$$\mathrm{Ric}_L(X) = \sum_{j=2}^{k} K(X \wedge e_j). \tag{7}$$

We note that such a curvature is called a *k*-Ricci curvature.

For an integer *k*, 2≤k≤n, B.-Y. Chen introduced a Riemannian invariant Θk defined by

$$\Theta_k(p) = \frac{1}{k-1}\inf\left\{\mathrm{Ric}_L(X)|L,X\right\}, \quad p \in M, \tag{8}$$

where *L* runs over all *k*-plane sections in TpM and *X* runs over all unit vectors in *L*.

It is well known that all these above invariants have many interesting applications to several fields of mathematics (see [30]).

2.2 Quaternionic Kähler manifolds

The geometry of Riemannian structures, complex structures, almost contact structures, hypercomplex structures, quaternionic and Cauchy-Riemann structures belongs to the general theory of the *G*-structures, a domain of great interest in modern differential geometry, in global analysis and mathematical physics. Quaternionic manifolds correspond to the reduction of the structural group at GL(n,H)·Sp(1)$GL(n,H)\cdot Sp(1)$. From the metric viewpoint, the most interesting is the case of quaternionic Kähler manifolds, which corresponds to the reduction of the holonomy at a subgroup of Sp(n)·Sp(1)$Sp(n)\cdot Sp(1)$. They appear in Berger's list for possible holonomy groups of irreducible Riemannian manifolds [31]. We give in this subsection a quick review of basic definitions and properties concerning the differential geometry of manifolds endowed with quaternionic structures. For details, see [32].

Let $\overline{M}$ be a differentiable manifold and assume that there is a rank 3-subbundle *σ* of End $(T\overline{M})$such that a local basis {J1,J2,J3} exists on sections of *σ* satisfying for all α∈{1,2,3}:

$$J_\alpha^2 = -\mathrm{Id}, \qquad J_\alpha J_{\alpha+1} = -J_{\alpha+1}J_\alpha = J_{\alpha+2}, \tag{9}$$

where Id denotes the identity tensor field of type (1,1)$(1,1)$ on M and the indices are taken from {1,2,3} modulo 3. Then the bundleσ is called an almost quaternionic structure on M and {J1,J2,J3} is called a canonical local basis of σ. Moreover, $(\overline{M},\sigma)$ is said to be an almost quaternionic manifold. It is easy to see that any almost quaternionic manifold is of dimension $4m$, m≥1.

A Riemannian metric $\overline{g}$ on $\overline{M}$is said to be adapted to the almost quaternionic structure σ if it satisfies:

$$\overline{g}(J_\alpha X, J_\alpha Y) = \overline{g}(X, Y), \quad \forall \alpha \in \{1,2,3\} \tag{10}$$

for all vector fields X, Y on $\overline{M}$ and any canonical local basis {J1,J2,J3} of σ. Moreover, $(\overline{M},\sigma,\overline{g})$is said to be an almost quaternionic Hermitian manifold.

If the bundle σ is parallel with respect to the Levi-Civita connection $\overline{\nabla}$ of $\overline{g}$, then $(\overline{M},\sigma,\overline{g})$ is said to be a quaternionic Kähler manifold. Equivalently, locally defined 1-forms ω1ω_1, ω2ω_2, ω3ω_3 exist such that we have, for all $\alpha \in \{1,2,3\}$,

$$\overline{\nabla}_X J_\alpha = \omega_{\alpha+2}(X) J_{\alpha+1} - \omega_{\alpha+1}(X) J_{\alpha+2} \tag{11}$$

for any vector field X on $\overline{M}$, where the indices are taken from $\{1,2,3\}$ modulo 3.

We remark that any quaternionic Kähler manifold is an Einstein manifold, provided that dimM>4 (see [32–34]).

Let$(\overline{M},\sigma,\overline{g})$ be a quaternionic Kähler manifold and let X be a non-null vector on $\overline{M}$. Then the 4-plane spanned by {X,J1X,J2X,J3X}, denoted by Q(X), is called a quaternionic 4-plane. Any 2-plane in Q(X) is called a quaternionic plane. The sectional curvature of a quaternionic plane is called a quaternionic sectional curvature. A quaternionic Kähler manifold is a quaternionic space form if its quaternionic sectional curvatures are equal to a constant, say c. It is well-known that a quaternionic Kähler manifold $(\overline{M},\sigma,\overline{g})$ is a quaternionic space form, denoted $\overline{M}(c)$, if and only if its curvature tensor is given by (see [32])

$$\overline{R}(X,Y)Z = \frac{c}{4}\left\{\overline{g}(Z,Y)X - \overline{g}(X,Z)Y + \sum_{\alpha=1}^{3}\left[\overline{g}(Z,J_\alpha Y)J_\alpha X - \overline{g}(Z,J_\alpha X)J_\alpha Y + 2\overline{g}(X,J_\alpha Y)J_\alpha Z\right]\right\} \tag{12}$$

for all vector fields *X, Y, Z* on $\overline{M}$ and any local basis $\{J_1,J_2,J_3\}$ of σ. We note that some interesting characterizations of quaternionic space forms were obtained in [35].

For a submanifold *M* of a quaternion Kähler manifold $(\overline{M},\sigma,\overline{g})$, we denote by *g* the metric tensor induced on *M*. If ∇ is the covariant differentiation induced on *M*, the Gauss and Weingarten formulas are given by

$$\overline{\nabla}_X Y = \nabla_X Y + h(X,Y), \quad \forall X,Y \in \Gamma(TM) \tag{13}$$

and

$$\overline{\nabla}_X N = -A_N X + \nabla_X^{\perp} N, \quad \forall X \in \Gamma(TM), \forall N \in \Gamma\left(TM^{\perp}\right), \tag{14}$$

where *h* is the second fundamental form of *M*, ∇⊥$\nabla^{\perp}$ is the connection on the normal bundle and ANA_N is the shape operator of *M*with respect to *N*. The shape operator ANA_N is related to *h* by

$$g(A_N X,Y) = \overline{g}\left(h(X,Y),N\right)$$

for all X,Y∈Γ(TM) and N∈Γ(TM⊥).

If {e1,...,en} is an orthonormal basis oT_pM and {en+1,...,e4m} is an orthonormal basis of $T_p^{\perp}M$, where$p \in M$, we denote by *H* the mean curvature vector, that is,

$$H(p) = \frac{1}{n}\sum_{i=1}^{n} h(e_i, e_i).$$

Also, we set

$$h_{ij}^r = g\big(h(e_i, e_j), e_r\big), \quad i, j \in \{1, \ldots, n\}, r \in \{n+1, \ldots, 4m\}$$

and

$$\|h\|^2(p) = \sum_{i,j=1}^{n} g\big(h(e_i, e_j), h(e_i, e_j)\big).$$

The submanifold *M* is called totally geodesic if the second fundamental form vanishes identically and totally umbilical if there is a real number λ such that h(X,Y)=λg(X,Y)H for any tangent vectors *X*, *Y* on *M*. If H=0, then the submanifold *M* is said to be minimal.

3 FUNDAMENTAL INEQUALITIES INVOLVING CHEN INVARIANTS AND THE SQUARED MEAN CURVATURE

For a Riemannian submanifold M^n of a real space form $\overline{M}$ with constant sectional curvature *c*, Chen [1] proved the following inequality for the Riemannian invariant δ_M of *M*, known as the first Chen inequality:

$$\delta_M \leq \frac{n-2}{2}\left[\frac{n^2}{n-1}\|H\|^2 + c(n+1)\right]. \tag{15}$$

The submanifold *M* is said to satisfy Chen's equality if the equality case of (15) holds identically.

In quaternionic Kähler ambient, the first classes of submanifolds which have been introduced and studied were quaternionic [36] and totally real submanifolds [37]. A submanifold M in a quaternionic Kähler manifold $(\overline{M}, \sigma, \overline{g})$ is called a quaternionic submanifold (resp. a totally real submanifold) if each tangent space of M is carried into itself (resp. into the normal space) by each section in σ. An n-dimensional totally real submanifold of a quaternionic space form $\overline{M}^{4m}(c)$ is said to be a Lagrangian submanifold if n=m. It is well known that a quaternionic submanifold is totally geodesic, so the first class of interest from Chen's inequalities viewpoint is given by the totally real submanifolds. In [38], the validity of the inequality (15) was proved for totally real submanifolds in a quaternionic space form $\overline{M}(4c)$ of quaternionic sectional curvature $4c$, and Chen's equality was interpreted in terms of eigenvalues and eigenspaces of the Weingarten operators of the submanifold. As a consequence, Hong and Houh obtained the following interesting result.

Theorem 3.1 [38]

Every Lagrangian submanifold of a quaternionic space form satisfying Chen's equality is minimal.

In [39], Şahin introduced the concept of slant submanifolds as a natural generalization of both quaternionic and totally real submanifolds. A submanifold M of a quaternionic Kähler manifold $\overline{M}$ is said to be a slant submanifold if for each non-zero vector$X$ tangent to M at p, the angle $\theta(X)$ between $J_\alpha(X)$ and T_pM, $\alpha\in\{1,2,3\}$ is constant, *i.e.*, it does not depend on choice of $p\in M$ and $X\in T_pM$. We can easily see that quaternionic submanifolds are slant submanifolds with $\theta=0$ and totally-real submanifolds are slant submanifolds with $\theta = \frac{\pi}{2}$. A slant submanifold of a quaternionic Käler manifold is said to be proper (or θ-slant proper) if it is neither quaternionic nor totally real.

In [40], the present author obtained the generalization of the first Chen inequality to the case of slant submanifolds in quaternionic space forms as follows.

Theorem 3.2 [40]

Let Mn *be a θ-slant proper submanifold of a quaternionic space form* $\overline{M}^{4m}(c)$*Then, for each point* p∈M, *we have*

$$\delta_M(p) \leq \frac{n-2}{2}\left[\frac{n^2}{n-1}\|H\|^2 + \frac{c}{4}\left(n+1+9\cos^2\theta\right)\right]. \tag{16}$$

The equality in (16) *holds at* p∈M *if and only if there exists an orthonormal basis* {e1,...,en} *of* TpM *and an orthonormal basis* {en+1,...,e4m} *of* $T_p^{\perp}M$ *such that the shape operators* Ar≡Aer, r∈{n+1,...,4m}, *take the following forms*:

$$A_{n+1} = \begin{pmatrix} a & 0 & 0 & \cdots & 0 \\ 0 & a & 0 & \cdots & 0 \\ 0 & 0 & a+b & \cdots & 0 \\ \vdots & \vdots & \vdots & \ddots & \vdots \\ 0 & 0 & 0 & \cdots & a+b \end{pmatrix} \tag{17}$$

and

$$A_r = \begin{pmatrix} a_r & b_r & 0 & \cdots & 0 \\ a_r & -b_r & 0 & \cdots & 0 \\ 0 & 0 & 0 & \cdots & 0 \\ \vdots & \vdots & \vdots & \ddots & \vdots \\ 0 & 0 & 0 & \cdots & 0 \end{pmatrix}, \quad r \in \{n+2, \ldots, 4m\}. \tag{18}$$

The second Chen fundamental inequality, stated in [41], asserts that for any submanifold MnM^n of a real space form $\overline{M}$ with constant sectional curvature c, we have

$$\delta(n_1, \ldots, n_k) \leq d(n_1, \ldots, n_k)\|H\|^2 + b(n_1, \ldots, n_k)c \tag{19}$$

for any *k*-tuples (n1,...,nk)∈S(n). Immersion of a submanifold which realizes equality in this inequality at every point is said to be an ideal immersion.

In [42], Yoon proved that the inequality (19) is also true for totally real submanifolds in a quaternionic space forms of quaternionic sectional curvature 4*c*. This inequality was later generalized for slant submanifolds in quaternionic space forms as follows.

Theorem 3.3 [40]

If Mn *is a θ-slant proper submanifold of a quaternionic space form* $\overline{M}^{4m}(c)$, *then we have*

$$\delta(n_1,\ldots,n_k) \leq d(n_1,\ldots,n_k)\|H\|^2 + b(n_1,\ldots,n_k)\frac{c}{4} + \frac{9c}{8}\left(n - \sum_{j=1}^{k} n_j\right)\cos^2\theta \tag{20}$$

for any k-tuples (n1,...,nk)∈S(n).

The equality in (20) *holds at* p∈M *if and only if there exists an orthonormal basis* {e1,...,en} *of* TpM *and an orthonormal basis* {en+1,...,e4m} *of* $T_p^{\perp}M$ *such that the shape operators* Ar≡Aer, r∈{n+1,...,4m}, *take the following forms*:

$$A_{n+1} = \begin{pmatrix} a_1 & 0 & 0 & \cdots & 0 \\ 0 & a_2 & 0 & \cdots & 0 \\ 0 & 0 & a_3 & \cdots & 0 \\ \vdots & \vdots & \vdots & \ddots & \vdots \\ 0 & 0 & 0 & \cdots & a_n \end{pmatrix} \tag{21}$$

and

$$A_r = \begin{pmatrix} B_1^r & \cdots & 0 & 0 & \cdots & 0 \\ \vdots & \ddots & \vdots & \vdots & \ddots & \vdots \\ 0 & \cdots & B_k^r & 0 & \cdots & 0 \\ 0 & \cdots & 0 & c_r & \cdots & 0 \\ \vdots & \ddots & \vdots & \vdots & \ddots & \vdots \\ 0 & \cdots & 0 & 0 & \cdots & c_r \end{pmatrix}, \quad r \in \{n+2, \ldots, 4m\}, \tag{22}$$

where a1,...,an *satisfy*

$$a_1 + \cdots + a_{n_1} = \cdots = a_{n_1+\cdots+n_{k-1}+1} + \cdots + a_{n_1+\cdots+n_k} = a_{n_1+\cdots+n_k+1} = \cdots = a_n$$

and each B_j^r *is a symmetric* nj×nj$n_j \times n_j$ *submatrix satisfying*:

$$\operatorname{trace}(B_1^r) = \cdots = \operatorname{trace}(B_k^r) = c_r.$$

4 FUNDAMENTAL INEQUALITIES INVOLVING THE RICCI CURVATURE AND THE SQUARED MEAN CURVATURE

In [4], Chen established a sharp relationship involving the Ricci curvature and the squared mean curvature for an arbitrary *n*-dimensional Riemannian submanifold of a real space form of constant sectional curvature *c*,

$$\operatorname{Ric}(X) \le (n-1)c + \frac{n^2}{4}\|H\|^2, \tag{23}$$

which is known as the Chen-Ricci inequality. In [43], Liu obtained the same inequality for totally real submanifolds in quaternionic space forms.

A submanifold *M* of a quaternion Kähler manifold$(\overline{M}, \sigma, \overline{g})$ is said to be a quaternionic CR-submanifold if there exists two orthogonal complementary distributions D and D⊥$D^{\perp}$ on M such that D is invariant under quaternionic structure and D⊥ is totally real (see [44]). An estimation of the Ricci curvature of a quaternioni CR-submanifold in a quaternionic space form has been established in [45], as follows.

Theorem 4.1 [45]

Let M be an n-dimensional quaternionic CR-submanifold of a quaternionic space form $\overline{M}(c)$*. Then:*

(i$_1$) *For each unit vector* $X \in D_p^{\perp}$, *we have*

$$\operatorname{Ric}(X) \leq \frac{(n-1)c}{4} + \frac{n^2}{4}\|H\|^2. \tag{24}$$

(i$_2$) *For each unit vector* X∈Dp, *we have*

$$\operatorname{Ric}(X) \leq \frac{(n+8)c}{4} + \frac{n^2}{4}\|H\|^2. \tag{25}$$

(ii)If H(p)=0, *then a unit tangent vector X at p satisfies the equality case of* (24) (*respectively* (25)) *if and only if* $X \in D_p^{\perp} \cap \mathcal{N}_p$ (*respectively* X∈Dp∩Np), *where* Np *is the relative null space of M at the point* p∈M *defined by*

$$\mathcal{N}_p = \{Z \in T_pM \mid h(Z, Y) = 0, \forall Y \in T_pM\}.$$

We note that an optimal inequality concerning the Ricci curvature for quaternionic CR-submanifolds of quaternionic space forms with a semi-symmetric metric connection was obtained in [46]. It is clear that, although quaternionic CR-submanifolds are also the generalization of both quaternionic and totally real submanifolds, there exists no inclusion between the two classes of quaternionic CR-submanifolds and slant submanifolds. The following estimation of the Ricci curvature for slant submanifolds in quaternionic space forms was firstly proved in [40], and an alternative nice proof can be found in [47].

Theorem 4.2 [40]

Let MnM^n *be a θ-slant proper submanifold of a quaternionic space form* $\overline{M}^{4m}(c)$. *Then*:

(i)For each unit vector X∈TpM, *we have*

$$\mathrm{Ric}(X) \le \frac{(n-1)c}{4} + \frac{n^2}{4}\|H\|^2 + \frac{3c}{8}\cos^2\theta. \tag{26}$$

(ii)If H(p)=0, *then a unit tangent vector X at p satisfies the equality case of* (26) *if and only if X belongs to the relative null space of M at p.*

(iii)The equality case of (26) *holds identically for all unit tangent vectors at p if and only either p is a totally geodesic point or* n=2$n=2$ *and p is a totally umbilical point.*

In [48], Oprea proved using optimization methods that the inequality (23) is not optimal for Lagrangian submanifolds in complex space forms and obtained an improved Chen-Ricci inequality. In [49], Deng obtained the proof of the improved inequality as an application of suitable algebraic inequalities. This allows one to determine the equality condition in an explicit form. Recently, in [50], the Chen-Ricci inequality was improved for Lagrangian submanifolds in quaternionic space forms as follows.

Theorem 4.3 [50]

Let Mn *be a Lagrangian submanifold of real dimension* n≥2 *in a quaternionic space form* $\overline{M}^{4n}(c)$, *p be a point in M and X be a unit tangent vector in* TpM. *Then we have*

$$\mathrm{Ric}(X) \le \frac{(n-1)}{4}\left(c + \|H\|^2\right). \tag{27}$$

Moreover, the equality sign holds for any unit tangent vector at p if and only if either

(i)p is a totally geodesic point or

*(ii)*n=2 *and p is an H-umbilical point with* λr=3μr, r=1,2,3, *i.e., there exists an orthonormal basis* {e1,...,en} *of* TpM *such that the second fundamental form takes the following simple form*:

$$h(e_1, e_1) = \sum_{\alpha=1}^{3} \lambda_\alpha J_\alpha(e_1),$$

$$h(e_2, e_2) = \sum_{\alpha=1}^{3} \mu_\alpha J_\alpha(e_1),$$

$$h(e_1, e_j) = \sum_{\alpha=1}^{3} \mu_\alpha J_\alpha(e_j),$$

$$h(e_j, e_k) = 0, \quad j \neq k, j, k = 2, \ldots, n$$

for some suitable functions λrλ_r *and* μr *satisfying* λr=3μr, r=1,2,3.

5 FUNDAMENTAL INEQUALITIES INVOLVING K-RICCI CURVATURE, SQUARED MEAN CURVATURE AND SHAPE OPERATOR

In [4], Chen extended the notion of Ricci curvature to *k*-Ricci curvature for a Riemannian manifold and established a sharp relationship between *k*-Ricci curvatures and the shape operator and also a sharp relationship between *k*-Ricci curvatures and the squared mean curvature for an *n*-dimensional Riemannian submanifold in a real space form $\overline{M}$ with constant sectional curvature*c*. More precisely, he proved that for any point p∈M and any integer *k*, 2≤k≤n, one has

$$\|H\|^2(p) \geq \frac{4(n-1)}{n^2}\left[\frac{\Theta_k(p)}{k-1} - c\right]. \tag{28}$$

On the other hand, Chen proved that if Ɵk(p)≠c, then the shape operator at the mean curvature vector satisfies for any integer*k*, 2≤k≤n, and any point p∈M,

$$A_H > \frac{n-1}{n}\left[\Theta_k(p) - c\right]I_n, \tag{29}$$

at *p*, where InI_n denotes the identity map of TpMn.

The inequality (28) was extended in the setting of a totally real submanifold in a quaternionic space form by Liu and Dai in [51] and recently, in [52], both inequalities (28) and (29) were generalized for proper slant submanifolds in quaternionic space forms as follows.

Theorem 5.1 [52]

Let MnM^n *be a θ-slant proper submanifold of a quaternionic space form* $\overline{M}^{4m}(c)$. *Then, for any* p∈M$p \in M$ *and any integer k,* 2≤k≤n, *one has*

$$\|H\|^2(p) \geq \Theta_k(p) - \frac{c}{4}\left(1 + \frac{9}{n-1}\cos^2\theta\right). \tag{30}$$

Theorem 5.2 [52]

Let x: $M \rightarrow \overline{M}^{4m}(c)$ *be an isometric immersion of an n-dimensional θ-slant proper submanifold M into a 4m-dimensional quaternionic space form*$\overline{M}(c)$. *Then, for any* p∈M *and any integer k,* 2≤k≤n, *one has*:

(i)If $\Theta_k(p) \neq \frac{c}{4}(1 + \frac{9}{n-1}\cos^2\theta)$, *then the shape operator at the mean curvature satisfies*

$$A_H > \frac{n-1}{n}\left[\Theta_k(p) - \frac{c}{4}\left(1 + \frac{9}{n-1}\cos^2\theta\right)\right]I_n, \tag{31}$$

at p, where InI_n *denotes the identity map of* TpM.

(ii)If $\Theta_k(p) = \frac{c}{4}(1 + \frac{9}{n-1}\cos^2\theta)$*, then* AH≥0 *at p.*

(iii)A unit vector X∈TpM *satisfies*

$$A_H X = \frac{n-1}{n}\left[\Theta_k(p) - \frac{c}{4}\left(1 + \frac{9}{n-1}\cos^2\theta\right)\right]X \tag{32}$$

if and only if $\Theta_k(p) = \frac{c}{4}(1 + \frac{9}{n-1}\cos^2\theta)$ *and X belongs to* Np, *the relative null space of M at p.*

(iv) The identity

$$A_H = \frac{n-1}{n}\left[\Theta_k(p) - \frac{c}{4}\left(1 + \frac{9}{n-1}\cos^2\theta\right)\right]I_n \tag{33}$$

holds at p if and only if p is a totally geodesic point.

6 SOME OPEN PROBLEMS

In this section we propose to investigate the following open problems related to the Chen invariants and ideal immersions in quaternionic space forms.

Problem 6.1 QR-submanifolds were introduced by Bejancu [53] as a generalization of the real hypersurfaces of a quaternionic Kähler manifold. In fact, a real submanifold *M* of a quaternionic Kähler manifold $(\overline{M}, \sigma, \overline{g})$ is called a QR-submanifold if there exists a vector subbundle *D* of the normal bundle TM⊥ such that Jα(Dp)=Dp and Jα(D⊥p)⊂TpM, for all p∈M, α=1,2,3$\alpha = 1,2,3$ and for any local basis {J1,J2,J3} of *σ*, where D⊥ is the complementary orthogonal bundle to *D* in TM⊥. The problem is to extend the inequalities (15), (19), (23), (28) and (29) for QR-submanifolds in quaternionic space forms.

Problem 6.2 To generalize the Theorems 3.2, 3.3, 4.1, 4.2, 5.1 and 5.2 for semi-slant submanifolds in quaternionic space forms.

We note that the concept of a semi-slant submanifold in quaternionic geometry was introduced by Şahin [39] as follows: a real submanifold *M* of a quaternionic Kähler manifold (M,σ,g¯) is said to be a semi-slant submanifold if there exist two orthogonal vector subbundles μ and μ⊥ of the normal bundle TM⊥ such that $TM^{\perp} = \mu \oplus \mu^{\perp}$, $\mu_p^{\perp}$ is anti-invariant with respect to Jα$J_\alpha$and μp is slant with respect to JαJ_α for all p∈M, α=1,2,3 and for any local basis {J1,J2,J3} of σ. It is easy to see that this notion is natural because QR-submanifolds and in particular real hypersurfaces are examples of semi-slant submanifolds of a quaternionic Kähler manifold and therefore fulfill the main purpose for which semi-slant submanifolds were introduced in Kähler geometry by Papaghiuc [54]. We also remark that recently, in [55], Shukla and Rao defined another concept of a semi-slant submanifold of a quaternionic Kähler manifold by analogy with the definition of Papaghiuc, but that class of submanifolds, although generalizes slant submanifolds, does not contain real hypersurfaces as a subclass.

Problem 6.3 Recently, Tripathi [56], Mihai and Rădulescu [57] obtained an improved Chen-Ricci inequality for Kählerian slant submanifolds in a complex space form. On the other hand, the quaternionic version of a Kählerian slant submanifold has been introduced in [39], under the name of a quaternionic slant submanifold, and some properties were obtained in [52]. Thus a proper slant submanifold *M* of a quaternionic Kähler manifold $(\overline{M},\sigma,\overline{g})$ is said to be a quaternionic slant submanifold if it satisfies the condition

$$(\overline{\nabla}_X P_\alpha)Y = \omega_{\alpha+2}(X)P_{\alpha+1}Y - \omega_{\alpha+1}(X)P_{\alpha+2}Y$$

for all vector fields *X*, *Y* on $\overline{M}$, where the indices are taken from {1,2,3} modulo 3 and PαY denotes the tangential component of JαY. The problem is to extend the improved Chen-Ricci inequality (27) to quaternionic slant submanifolds in quaternionic space forms and to investigate the equality case of the inequality.

Problem 6.4 In [58], Oprea improved the inequality (15) for Lagrangian submanifolds in complex space forms and recently, in [59], Chen and Dillen obtained improved general inequalities which involve the squared mean curvature and the invariants δ(n1,...,nk) for Lagrangian submanifolds in complex space forms, giving also necessary and sufficient condition for a Lagrangian submanifold to attain the equality for arbitrary δ(n1,...,nk). The problem is to obtain improved general inequalities for Lagrangian submanifolds in quaternionic space forms and to completely classify Lagrangian submanifolds which realize the equality case of these inequalities.

Problem 6.5 In [52] it was proved that there do not exist quaternionic slant immersions of minimal codimension in quaternionic projective space with unfull first normal bundle. An interesting problem is to investigate the existence of quaternionic slant immersions of minimal codimension in a quaternionic projective space Pn(H) satisfying the equality case of (19) such that eithern1+n2+⋯+nk=n or n1+n2+⋯+nk<n and at least one of nin_i is 2. We note that the answer of the corresponding problem in Kählerian geometry is negative [60].

Problem 6.6 To completely classify δ(2) -ideal slant submanifolds and δ(2,2) -ideal Lagrangian submanifolds in quaternionic space forms.

We note that δ(2) and δ(2,2) are the simplest non-trivial δ-curvature invariants and some classification results for δ(2) and δ(2,2) -ideal Lagrangian submanifolds in complex space forms were obtained in [61–64].

Problem 6.7 To obtain Chen-like inequalities for the Casorati curvatures of slant submanifolds in quaternionic space forms and to completely classify Casorati ideal submanifolds.

It is known that the Casorati curvature of a submanifold in a Riemannian manifold is an extrinsic invariant defined as the normalized square of the length of the second fundamental form. Moreover, this notion extends the concept of the principal direction of a hypersurface of a Riemannian manifold to submanifolds of a Riemannian manifold and it was preferred by Casorati over the traditional Gauss curvature because corresponds better with the common intuition of curvature [65]. Therefore it is of great interest to obtain optimal inequalities for the Casorati curvatures of submanifolds in different ambient spaces. We note that in [66], Decu, Haesen and Verstraelen obtained some optimal inequalities involving the scalar curvature and the Casorati curvature of a Riemannian submanifold in a real space form and the holomorphic sectional curvature and the Casorati curvature of a Kähler hypersurface in a complex space form. Moreover, the same authors proved in [67] an inequality in which the scalar curvature is estimated from above by the normalized Casorati curvatures.

ADDITIONAL INFORMATION

This paper is a contribution to the *International Conference on Applied Analysis and Algebra* (ICAAA2012) (20-24 June 2012, Istanbul, Turkey).

ACKNOWLEDGEMENTS

Dedicated to Prof. Ravi P Agarwal on his 65th birth anniversary.

This work was supported by CNCS-UEFISCDI, project number PN-II-ID-PCE-2011-3-0118.

COMPETING INTERESTS

The author declares that he has no competing interests.

REFERENCES

1. Chen B-Y: Some pinching and classification theorems for minimal submanifolds. *Arch. Math.* 1993, 60: 568–578.

2. Chern SS: *Minimal Submanifolds in a Riemannian Manifold.* University of Kansas Press, Lawrence; 1968.

3. Suceavă B, Vâjiac M: Remarks on Chen's fundamental inequality with classical curvature invariants in Riemannian spaces. *An. Ştiint. Univ. Al. I. Cuza Iaşi. Mat.* 2008, 54(1):27–37.

4. Chen B-Y: Relations between Ricci curvature and shape operator for submanifolds with arbitrary codimensions. *Glasg. Math. J.* 1999, 41: 33–41.

5. Chen B-Y: A general inequality for submanifolds in complex-space-forms and its applications. *Arch. Math.* 1996, 67(6):519–528.

6. Chen B-Y: An optimal inequality for CR-warped products in complex space forms involving CR δ -invariant. *Int. J. Math.* 2012., 23(3): Article ID 1250045

7. Oiagă A, Mihai I: B.-Y. Chen inequalities for slant submanifolds in complex space forms. *Demonstr. Math.* 1999, 32: 835–846.

8. Kim J-S, Choi J: A basic inequality for submanifolds in a cosymplectic space form. *Int. J. Math. Math. Sci.* 2003, 9: 539–547.

9. Liu X, Su W: Shape operator of slant submanifolds in cosymplectic space forms. *Studia Sci. Math. Hung.* 2005, 42(4):387–400.

10. Yoon DW: Inequality for Ricci curvature of slant submanifolds in cosymplectic space forms. *Turk. J. Math.* 2006, 30: 43–56.

11. Cioroboiu D, Oiagă A: B.Y. Chen inequalities for slant submanifolds in Sasakian space forms. *Rend. Circ. Mat. Palermo* 2003, 52: 367–381.

12. Defever F, Mihai I, Verstraelen L: B.-Y. Chen's inequality for *C* -totally real submanifolds of Sasakian space forms. *Boll. Unione Mat. Ital, B*1997, 11(2):365–374.

13. Mihai I: Ricci curvature of submanifolds in Sasakian space forms. *J. Aust. Math. Soc.* 2002, 72(2):247–256.

14. Carriazo A, Kim YH, Yoon DW: Some inequalities on totally real submanifolds in locally conformal Kähler space forms. *J. Korean Math. Soc.* 2004, 41(5):795–808.

15. Hong S, Matsumoto K, Tripathi MM: Certain basic inequalities for submanifolds of locally conformal Kähler space forms. *SUT J. Math.*2005, 41(1):75–94.

16. Kim J-S, Yoon DW: Inequality for totally real warped products in locally conformal Kaehler space forms. *Kyungpook Math. J.* 2004, 44(4):585–592.

17. Alegre P, Carriazo A, Kim YH, Yoon DW: B.-Y. Chen's inequality for submanifolds of generalized space forms. *Indian J. Pure Appl. Math.*2007, 38(3):185–201.

18. Kim J-S, Song YM, Tripathi MM: B.-Y. Chen inequalities for submanifolds in generalized complex space forms. *Bull. Korean Math. Soc.*2003, 40: 411–423.

19. Mihai A:Shape operator AHA_H for slant submanifolds in generalized complex space forms. *Turk. J. Math.* 2003, 27(4):509–523.

20. Arslan K, Ezentas R, Mihai I, Murathan C, Özgür C: B.Y. Chen inequalities for submanifolds in locally conformal almost cosymplectic manifolds. *Bull. Inst. Math. Acad. Sin.* 2001, 29(3):231–242.

21. Yoon DW: Inequality for Ricci curvature of certain submanifolds in locally conformal almost cosymplectic manifolds. *Int. J. Math. Math. Sci.*2005, 10: 1621–1632.

22. Arslan K, Ezentas R, Mihai I, Murathan C, Özgür C:Certain inequalities for submanifolds in (k,μ)(κ,μ)-contact space forms. *Bull. Aust. Math. Soc.* 2001, 64(2):201–212.

23. Tripathi MM, Kim JS: C -totally real submanifolds in (κ,μ)(κ,μ)-contact space forms. *Bull. Aust. Math. Soc.* 2003, 67(1):51–65.

24. Arslan K, Ezentas R, Mihai I, Murathan C, Özgür C: Ricci curvature of submanifolds in Kenmotsu space forms. *Int. J. Math. Math. Sci.*2002, 29(12):719–726.

25. Carriazo A, Fernandez LM, Hans-Uber MB: B.Y. Chen's inequality for S -space-forms: applications to slant immersions. *Indian J. Pure Appl. Math.* 2003, 34(9):1287–1298.

26. Fernández LM, Hans-Uber MB: New relationships involving the mean curvature of slant submanifolds in S -space-forms. *J. Korean Math. Soc.* 2007, 44(3):647–659.

27. Kim J-S, Dwivedi MK, Tripathi MM: Ricci curvature of integral submanifolds of an S -space form. *Bull. Korean Math. Soc.* 2007, 44(3):395–406.

28. Aktan N, Zeki Sarikaya M, Ozusaglam E: B.Y. Chen's inequality for semi-slant submanifolds in T-space forms. *Balk. J. Geom. Appl.* 2008, 13(1):1–10.

29. Chen B-Y Topics in Differential Geometry. In *δ-Invariants, Inequalities of Submanifolds and Their Applications.* Ed. Acad. Române, Bucharest; 2008:29–155.

30. Chen B-Y: *Pseudo-Riemannian Geometry, δ-Invariants and Applications.* World Scientific, Hackensack; 2011.

31. Berger M: Sur les groupes d'holonomie homogène des variétés à connexion affine et des variétés riemanniennes. *Bull. Soc. Math. Fr.*1955, 83: 279–330.

32. Ishihara S: Quaternion Kählerian manifolds. *J. Differ. Geom.* 1974, 9: 483–500.

33. Besse A: *Einstein Manifolds.* Springer, Berlin; 1987.

34. Salamon S: Differential geometry of quaternionic manifolds. *Ann. Sci. Éc. Norm. Super.* 1986, 19: 31–55.

35. Adachi T, Maeda S: Some characterizations of quaternionic space forms. *Proc. Jpn. Acad., Ser. A, Math. Sci.* 2000, 76(10):168–172.

36. Gray A:A note on manifolds whose holonomy group is a subgroup of Sp(n)·Sp(1)$Sp(n)\cdot Sp(1)$. *Mich. Math. J.* 1969, 16: 125–128.

37. Chen B-Y, Houh CS: Totally real submanifolds of a quaternion projective space. *Ann. Mat. Pura Appl.* 1979, 120(1):185–199.

38. Hong Y, Houh CS: Lagrangian submanifolds of quaternion Kaehlerian manifolds satisfying Chen's equality. *Beiträge Algebra Geom.* 1998, 39(2):413–421.

39. Şahin B: Slant submanifolds of quaternion Kaehler manifolds. *Commun. Korean Math. Soc.* 2007, 22(1):123–135.

40. Vîlcu GE: B.-Y. Chen inequalities for slant submanifolds in quaternionic space forms. *Turk. J. Math.* 2010, 34(1):115–128.

41. Chen B-Y: Some new obstructions to minimal and Lagrangian isometric immersions. *Jpn. J. Math.* 2000, 26(1):105–127.

42. Yoon DW: A basic inequality of submanifolds in quaternionic space forms. *Balk. J. Geom. Appl.* 2004, 9(2):92–102.

43. Liu X: On Ricci curvature of totally real submanifolds in a quaternion projective space. *Arch. Math.* 2002, 38(4):297–305.

44. Barros M, Chen B-Y, Urbano F: Quaternion CR-submanifolds of quaternion manifolds. *Kodai Math. J.* 1981, 4: 399–417.

45. Mihai I, Al-Solamy F, Shahid MH: On Ricci curvature of a quaternion CR-submanifold in a quaternion space form. *Rad. Mat.* 2003, 12(1):91–98.

46. Decu S: Optimal inequalities for submanifolds in quaternion-space-forms with semi-symmetric metric connection. *Bull. Transilv. Univ. Braşov, Ser. III* 2009, 2(51):175–184.

47. Shahid MH, Al-Solamy F: Ricci tensor of slant submanifolds in a quaternion projective space. *C. R. Math. Acad. Sci. Paris* 2011, 349(9):571–573.

48. Oprea, T: On a geometric inequality. arXiv:math.DG/0511088

49. Deng S: An improved Chen-Ricci inequality. *Int. Electron. J. Geom.* 2009, 2(2):39–45.

50. Deng S: Improved Chen-Ricci inequality for Lagrangian submanifolds in quaternion space forms. *Int. Electron. J. Geom.* 2012, 5(1):163–170.

51. Liu X, Dai W: Ricci curvature of submanifolds in a quaternion projective space. *Commun. Korean Math. Soc.* 2002, 17(4):625–633.

52. Vîlcu GE: Slant submanifolds of quaternionic space forms. *Publ. Math. (Debr.)* 2012, 81(3–4):397–413.

53. Bejancu A: QR-submanifolds of quaternion Kaehler manifolds. *Chin. J. Math.* 1986, 14: 81–94.

54. Papaghiuc N: Semi-slant submanifolds of a Kaehlerian manifold. *An. Ştiint. Univ. Al. I. Cuza Iaşi* 1994, 40(1):55–61.

55. Shukla SS, Rao PK: Ricci curvature of quaternion slant submanifolds in quaternion space forms. *Acta Math. Acad. Paedagog. Nyházi.*2012, 28(1):69–81.

56. Tripathi MM: Improved Chen-Ricci inequality for curvature-like tensors and its application. *Differ. Geom. Appl.* 2011, 29(5):685–698.

57. Mihai A, Rădulescu I: An improved Chen-Ricci inequality for Kaehlerian slant submanifolds in complex space forms. *Taiwan. J. Math.* 2012, 16(2):761–770.

58. Oprea T: Chen's inequality in the Lagrangian case. *Colloq. Math.* 2007, 108: 163–169.

59. Chen B-Y, Dillen F: Optimal general inequalities for Lagrangian submanifolds in complex space forms. *J. Math. Anal. Appl.* 2011, 379(1):229–239.

60. Li G, Wu C: Slant immersions of complex space forms and Chen's inequality. *Acta Math. Sci., Ser. B, Engl. Ed.* 2005, 25(2):223–232.

61. Chen B-Y, Dillen F, Verstraelen L, Vrancken L:Totally real submanifolds of CPnCP^n satisfying a basic equality. *Arch. Math.* 1994, 63(6):553–564.

62. Chen B-Y, Prieto-Martín A, Wang X:Lagrangian submanifolds in complex space forms satisfying an improved equality involving δ(2,2)$\delta(2,2)$.*Publ. Math. (Debr.)* 2013, 82(1):1–25.

63. Chen B-Y, Vrancken L: Lagrangian submanifolds satisfying a basic equality. *Math. Proc. Camb. Philos. Soc.* 1996, 120(2):291–307.

64. Chen B-Y, Vrancken L: Lagrangian submanifolds of the complex hyperbolic space. *Tsukuba J. Math.* 2002, 26(1):95–118.

65. Casorati F: Mesure de la courbure des surfaces suivant l'idée commune. Ses rapports avec les mesures de courbure gaussienne et moyenne. *Acta Math.* 1890, 14(1):95–110.

66. Decu S, Haesen S, Verstraelen L: Optimal inequalities involving Casorati curvatures. *Bull. Transilv. Univ. Braşov, Ser. B* 2007, 14(49):85–93. suppl.

67. Decu S, Haesen S, Verstraelen L: Optimal inequalities characterising quasi-umbilical submanifolds. *JIPAM. J. Inequal. Pure Appl. Math.*2008., 9(3): Article ID 79

5

NAVIER-STOKES EQUATIONS WITH VARIABLE VISCOSITY IN VARIABLE EXPONENT SPACES OF CLIFFORD-VALUED FUNCTIONS

Rui Niu[1,2], Hongtao Zheng[1] and Binlin Zhang[2]

[1]College of Power and Energy Engineering, Harbin Engineering University, Harbin, 150001, China.

[2]Department of Mathematics, Heilongjiang Institute of Technology, Harbin, 150050, China.

ABSTRACT

In this paper we study the stationary generalized Navier-Stokes equations when the viscosity is not only a constant but also a function which depends on the position and the shear-velocity. For this we establish an improved decomposition of variable exponent Lebesgue spaces of Clifford-valued functions. Using this decomposition together with Clifford operator calculus, we obtain the existence, uniqueness and representation of solutions for the generalized Stokes equations and the generalized Navier-Stokes equations with variable viscosity in the setting of variable exponent spaces of Clifford-valued functions. Furthermore, the equivalences of solutions and weak solutions for the aforementioned equations are justified.

KEYWORDS

Clifford analysis variable exponent, variable viscosity, Stokes equations, Navier-Stokes equations

1. INTRODUCTION

In this paper we are concerned with the stationary generalized Navier-Stokes equations:

$$-\operatorname{div}(A\nabla u) + \rho(u \cdot \nabla)u + \nabla q = \rho f \quad \text{in } \Omega, \tag{1.1}$$

$$\operatorname{div} u = 0 \quad \text{in } \Omega, \tag{1.2}$$

$$u = 0 \quad \text{on } \partial\Omega, \tag{1.3}$$

where the operator A is defined by $Au = au$ with $a(x,u) : \overline{\Omega} \times \mathbb{R}^n \to \mathbb{R}^+$ and $a \in C^\infty(\overline{\Omega} \times \mathbb{R}^n)$ and $\Omega \subset \mathbb{R}^n$ $(n \geq 2)$ and is a bounded domain with sufficiently smooth boundary $\partial\Omega$, u is the velocity, q the hydrostatic pressure, ρ the density, f the vector of the external forces. Notice that (1.1) may be written

$$-\frac{1}{\rho}\operatorname{div}(A(x,u)\mathrm{D}u) + (u \cdot \nabla)u + \nabla\pi = f \quad \text{in } \Omega, \tag{1.4}$$

where π:=q/ρπ:=q/ρ. If the relation $A(x,u) := \mu(\alpha + |\mathrm{D}u|^2)^{\frac{p-2}{2}}$ with μ,α>0μ,α>0, where μ is the *shear viscosity*, $\mathrm{D}u := (1/2)(\nabla u + (\nabla u)^\top)$ is the *symmetric gradient*, then the fluid is called *non-Newtonian fluid with p-structure*; see, for example, Acerbi and Mingione [1] for related discussions in this direction. Clearly, the fluid in the case p=2 is a *Newtonian fluid*, and then (1.4) becomes

$$-\nu\Delta u + (u \cdot \nabla)u + \nabla\pi = f \quad \text{in } \Omega, \tag{1.5}$$

where ν:=μ/ρν:=μ/ρ is the *kinematic viscosity*. Here (1.5) is the famous Navier-Stokes equation, for the detailed account about the Navier-Stokes equations we refer to [2, 3]. In this paper we will consider the Navier-Stokes equations in a Clifford language under the assumption that the viscosity depends on the position and velocity, *i.e.*, ν=ν(x,u)∈R+ν=ν(x,u)∈R+. It is easy to see that the viscous term −νΔu can be replaced by the Clifford expression DνDu, where D=∑nj=1ej∂j denotes the Dirac operator of a massless field, *u* is a Clifford-valued function; for the details as regards the Clifford algebra we refer to the next section.

In order to study the time-dependent motion of a viscous, incompressible fluid we need velocity fields *u* of the particles of the fluid along with their first spatial derivatives to be summable to the p(k)p(k)th power (p−≥1p−≥1, k∈Nk∈N) at each time and each position. Additionally, *u* has to be solenoidal and vanishes at the boundary of the domain where the motion happens. Thus this gives rise to the introduction of spaces of divergence free functions in a generalized sense. Furthermore, the models of electrorheological fluids, which were introduced by Rajagopal and Růžička [4, 5], can be described by the boundary value problems for the generalized Navier-Stokes equations. This leads naturally to the establishment of function spaces with variable exponents. On the other hand, Clifford algebras have important applications in a variety of fields including geometry, theoretical physics and digital image processing. They are named after the English geometer *William Kingdon Clifford*. We have a generalization of the complex numbers, the quaternions, and the exterior algebras; see [6]. As an active branch of mathematics over the past 40 years, Clifford analysis usually studies the solutions of the Dirac equation for functions defined on domains in Euclidean space and taking value in Clifford algebras; see, for example, [7]. In [8] the authors gave an overview of applications of Clifford analysis in mathematical physics. Hence, it makes sense to study the stationary Navier-Stokes equations in the setting of Clifford algebras.

It is worth pointing out that our attempt is to give a unified approach to deal with physical problems modeled by the generalized Navier-

Stokes equations, which is quite different from the approaches of some authors, for example, we refer the reader to the monograph [9]. Based on the above consideration, we should study the generalized Navier-Stokes equations in variable exponent spaces of Clifford-valued functions. Of course, the study of variable exponent spaces has been driven by various problems in elastic mechanics, calculus of variations and differential equations with variable growth; see [10–20] and the references therein.

Evidently, we should primarily be concerned with the study of the first term DA(Du)DA(Du) in (1.1). In [21, 22], Nolder first introduced the general nonlinear *A*-Dirac equations DA(x,Du)=0DA(x,Du)=0 which arise in the study of many phenomena in the physical sciences. In particular, he developed some tools for the study of weak solutions to nonlinear *A*-Dirac equations in the space W1,p0(Ω,Cℓn)W01,p(Ω,Cℓn). Inspired by his works, Fu and Zhang in [23, 24] were interested in the existence of weak solutions for the general nonlinear *A*-Dirac equations with variable growth. For this purpose, the authors established a theory of variable exponent spaces of Clifford-valued functions with applications to homogeneous and nonhomogeneous *A*-Dirac equations; see also [25, 26]. Very recently, Fu *et al.* in [27, 28] established a Hodge-type decomposition of variable exponent Lebesgue spaces of Clifford-valued functions with applications to the Stokes equations, the Navier-Stokes equations and the *A*-Dirac equations DA(Du)=0DA(Du)=0. By using the Hodge-type decomposition and variational methods, Molica Bisci *et al.* in [29] studied the properties of weak solutions to the homogeneous and nonhomogeneous *A*-Dirac equations with variable growth. For an overview for the nonlinear *A*-Dirac equations with variable growth, we refer the reader to [30].

It is worth mentioning that Diening *et al.* in [31] studied the following model introduced in [4, 5] to describe the motions of electrorheological fluids:

$$\begin{cases} -\operatorname{div}\mathcal{M}(\mathrm{D}u)+(u\cdot\nabla)u+\nabla\pi=f, & x\in\Omega,\\ \operatorname{div}u=0, & x\in\Omega,\\ u=0, & x\in\partial\Omega, \end{cases} \tag{1.6}$$

where $f\in(W_0^{1,p(x)}(\Omega))^*=W^{-1,p'(x)}(\Omega)$, $2n/(n+2)<p_-\leq p_+<\infty$ and the operator *A* satisfies certain natural variable growth conditions. The authors obtained the existence of weak solutions in $(W_0^{1,p(x)}(\Omega))^n\times L_0^s(\Omega)$, here $s:=\min\{(p_+)',np_-/2(n-p_-)\}$if p−<n and s:=(p+)′ otherwise, $L_0^s(\Omega):=\{\pi\in L^s(\Omega):\int_\Omega\pi\,dx=0\}$. From a practical point of view, we have to investigate the representation of solutions to system (1.6) besides existence and uniqueness. Based on the method developed by Sprößig in [2], it is possible to obtain the desired results if we consider system (1.6) as M(Du)=a(x,u)Du under the assumption $f\in W^{-1,p(x)}(\Omega,\mathbb{R}^n)$, here a(x,u)a(x,u) is a positive function.

Motivated by their works, our goal in this article is to give a generalization of related results in Sprößig [2] to the variable exponent setting. More precisely, we investigate properties of solutions for the following Navier-Stokes equations in variable exponent spaces of Clifford-valued functions:

$$\begin{cases} DADu+[uD]_0u+D\pi=f, & x\in\Omega,\\ [Du]_0=0, & x\in\Omega,\\ u=0, & x\in\partial\Omega. \end{cases} \tag{1.7}$$

Throughout the paper, the operator A is defined by Au=au *with* $a(x,u):\overline{\Omega}\times\mathbb{R}^n\to\mathbb{R}^+$*and* $a\in C^\infty(\overline{\Omega}\times\mathbb{R}^n)$.. The outline for this study is to first establish a modified decomposition of variable exponent Lebesgue spaces. Then the classic results about the Stokes equations and the Navier-Stokes equations obtained by Gürlebeck and Sprößig [32, 33] are extended to the variable exponent setting. In particular, we would like to point out the equivalences of solutions and weak solutions for the above-mentioned equations, which has not been clearly stated by the previous works in [28].

This paper is organized as follows. In Section 2, we start with a brief summary of Clifford algebra and variable exponent spaces of Clifford-valued functions. In Section 3, we establish a modified decomposition of variable exponent Lebesgue spaces, and then discuss its some applications, which will be needed later. In Section 4, we obtain the existence and uniqueness of the generalized Stokes equations in the context of variable exponent spaces. Moreover, the equivalence of solutions and weak solutions for the above-mentioned equations are showed. In Section 5, using similar methods to [28], we prove the existence and uniqueness of solutions to the generalized Navier-Stokes equations in $W_0^{1,p(x)}(\Omega, C\ell_n) \times L^{p(x)}(\Omega, \mathbb{R})$ under certain hypotheses. Moreover, the equivalence of solutions and weak solutions for the above-mentioned equations are presented.

2 PRELIMINARIES

First we recall some related definitions and results concerning Clifford algebra and variable exponent spaces of Clifford-valued functions. For a detailed account we refer to [6, 34, 35].

Let Cℓn be the real universal Clifford algebra over Rn. Denote Cℓn by

$$C\ell_n = \mathrm{span}\{e_0, e_1, e_2, \ldots, e_n, e_1e_2, \ldots, e_{n-1}e_n, \ldots, e_1e_2 \cdots e_n\},$$

where e0=1 (the identity element in Rn), {e1,e2,...,en} is an orthonormal basis of RnRn with the relation eiej+ejei=−2δije0. Thus the dimension of Cℓn is 2n. For I={i1,...,ir}⊂{1,...,n} with 1≤i1<⋯≤n1<i2, put eI=ei⋯eireI=ei1ei2⋯eir, while for I=∅I=∅, e∅=e0e∅=e0. For 0≤r≤n fixed, the space CℓrnCℓnr is defined by

$$C\ell_n^r = \text{span}\{e_I : |I| = r\},$$

where |I||I| denotes cardinal number of the set *I*. The Clifford algebra Cℓn is a graded algebra as

$$C\ell_n = \bigoplus_r C\ell_n^r.$$

Any element a∈Cℓna∈Cℓn may thus be written in a unique way as

$$a = [a]_0 + [a]_1 + \cdots + [a]_n,$$

where $[\,]_r : C\ell_n \to C\ell_n^r$ nr denotes the projection of $C\ell_n$ onto $C\ell_n^r$. In particular, by $C\ell_n^2 = \mathbb{H}$ we denote the algebra of real quaternions. It is customary to identify ℝ with Cℓ0nCℓn0 and identify Rn with Cℓ1n, respectively. This means that each element *x* of Rn may be represented by

$$x = \sum_{i=1}^{n} x_i e_i.$$

For u∈Cℓn, we denote by [u]0 the scalar part of *u*, that is, the coefficient of the element e0. We define the Clifford conjugation as follows:

$$\overline{e_{i_1} e_{i_2} \cdots e_{i_r}} = (-1)^{\frac{r(r+1)}{2}} e_{i_1} e_{i_2} \cdots e_{i_r}.$$

We denote

$$(A, B) = [\overline{A}B]_0.$$

Then an inner product is thus obtained, giving rise to the norm $|\cdot|$ on $C\ell_n$given by

$|A|^2 = [\overline{A}A]_0$.

A Clifford-valued function $u : \Omega \to C\ell_n$ can be written as $u = \sum_I u_I e_I$, where the coefficients uI:Ω→RuI:Ω→R are real-valued functions.

The Dirac operator on Euclidean space used here is introduced by

$$D = \sum_{j=1}^{n} e_j \partial_j.$$

This is a special case of the Atiyah-Singer-Dirac operator acting on sections of a spinor bundle. Note that the most famous Dirac operator describes the propagation of a free fermion in three dimensions.

If *u* is a real-valued function defined on a domain Ω in RnRn, then Du=∇u=(∂1u,∂2u,...,∂nu). Moreover, D2=DD=−Δ, where Δ is the Laplace operator which operates only on coefficients. A function is left monogenic if it satisfies the equation Du(x)=0 for each x∈Ω. A similar definition can be given for right monogenic function. An important example of a left monogenic function is the generalized Cauchy kernel

$$G(x) = \frac{1}{\omega_n} \frac{\overline{x}}{|x|^n},$$

where ωnωn denotes the surface area of the unit ball in RnRn. This function is a fundamental solution of the Dirac operator.

Next we recall some basic properties of variable exponent spaces of Clifford-valued functions. Note that in what follows, we use the short notation $L^{p(x)}(\Omega)$, $W^{1,p(x)}(\Omega)$, *etc.*, instead of $L^{p(x)}(\Omega,\mathbb{R})$, $W^{1,p(x)}(\Omega,\mathbb{R})$, *etc.* Throughout this paper we always assume (unless declare specially)

$$p \in \mathcal{P}^{\log}(\Omega) \quad \text{and} \quad 1 < p_- := \inf_{x \in \overline{\Omega}} p(x) \leq p(x) \leq \sup_{x \in \overline{\Omega}} p(x) =: p_+ < \infty, \tag{2.1}$$

where Plog(Ω) is the set of exponent *p* satisfying the so-called log-Hölder continuity, *i.e.*,

$$|p(x) - p(y)| \leq \frac{C}{\log(e + |x - y|^{-1})}$$

holds for all x,y∈Ω; see [36]. Let P(Ω) be the set of all Lebesgue measurable functions p:Ω→(1,∞). Given p∈P(Ω)we define the conjugate function p′(x)∈P(Ω)by

$$p'(x) = \frac{p(x)}{p(x) - 1}, \quad \forall x \in \Omega.$$

The variable exponent Lebesgue space $L^{p(x)}(\Omega)$ is defined by

$$L^{p(x)}(\Omega) = \left\{ u \in \mathcal{P}(\Omega) : \int_\Omega |u(x)|^{p(x)} dx < \infty \right\},$$

with the norm

$$\|u\|_{L^{p(x)}(\Omega)} = \inf\left\{ t > 0 : \int_\Omega \left| \frac{u(x)}{t} \right|^{p(x)} dx \leq 1 \right\}.$$

The variable exponent Sobolev space $W^{1,p(x)}(\Omega)$ is defined by

$$W^{1,p(x)}(\Omega) = \left\{ u \in L^{p(x)}(\Omega) : |\nabla u| \in L^{p(x)}(\Omega) \right\},$$

with the norm

$$\|u\|_{W^{1,p(x)}(\Omega)} = \|\nabla u\|_{L^{p(x)}(\Omega)} + \|u\|_{L^{p(x)}(\Omega)}. \tag{2.2}$$

Denote $W_0^{1,p(x)}(\Omega)$ by the completion of $C_0^{\infty}(\Omega)$ in $W_0^{1,p(x)}(\Omega)$ with respect to the norm (2.2). The space $W_0^{1,p(x)}(\Omega)$ is defined as the dual of the space $W_0^{1,p(x)}(\Omega)$. For more details we refer to [9] and references therein.

In what follows, we say $u \in L^{p(x)}(\Omega, C\ell_n)$ can be understood coordinate wisely. For example, $u \in L^{p(x)}(\Omega, C\ell_n)$ means that $\{u_I\} \subset L^{p(x)}(\Omega)$ for $u = \sum_I u_I e_I \in C\ell_n$ with the norm $\|u\|_{L^{p(x)}(\Omega, C\ell_n)} = \sum_I \|u_I\|_{L^{p(x)}(\Omega)}$. In this way, spaces $W^{1,p(x)}(\Omega, C\ell_n)$, $W_0^{1,p(x)}(\Omega, C\ell_n)$, $C_0^{\infty}(\Omega, C\ell_n)$, *etc.* can be understood similarly. In particular, the space $L^2(\Omega, C\ell_n)$ can be converted into a right Hilbert $C\ell_n$ -module by defining the following Clifford-valued inner product (see [32, Definition 3.74]):

$$(f, g)_{C\ell_n} = \int_{\Omega} \overline{f(x)} g(x)\, dx. \tag{2.3}$$

Remark 2.1

A simple calculation leads to the claim that the norm $\|u\|_{L^{p(x)}(\Omega, C\ell_n)}$) is equivalent to the norm $\||u|\|_{L^{p(x)}(\Omega)}$. Furthermore, $\|Du\|_{L^{p(x)}(\Omega, C\ell_n)}$ is an equivalent norm of $\|u\|_{W_0^{1,p(x)}(\Omega, C\ell_n)}$ for every $u \in W_0^{1,p(x)}(\Omega, C\ell_n)$; for the details we refer to [24, 25].

Lemma 2.1

(see [23, 24])

If $p(x) \in \mathcal{P}(\Omega)$, then $L^{p(x)}(\Omega, C\ell_n)$ and $W^{1,p(x)}(\Omega, C\ell_n)$ are reflexive Banach spaces.

Definition 2.1

(see [32])

Let u∈C(Ω,Cℓn). The Teodorescu operator is defined by

$$Tu(x) = \int_{\Omega} G(x-y)u(y)\,dy,$$

where G(x) is the generalized Cauchy kernel mentioned above.

Lemma 2.2 **(see [28]) The following operators are bounded linear operators*:***

(i) $T : L^{p(x)}(\Omega, C\ell_n) \to W^{1,p(x)}(\Omega, C\ell_n).$

(ii) $\widetilde{T} : W^{-1,p(x)}(\Omega, C\ell_n) \to L^{p(x)}(\Omega, C\ell_n),$ *where the operator* T˜T∼ *can be considered as a unique bounded linear extension of the* Teodorescu *operator.*

Lemma 2.3 ***(see [28])*** **The following operators are bounded linear operators*:***

(i) $D : W^{1,p(x)}(\Omega, C\ell_n) \to L^{p(x)}(\Omega, C\ell_n).$

(ii) $\widetilde{D} : L^{p(x)}(\Omega, C\ell_n) \to W^{-1,p(x)}(\Omega, C\ell_n),$*where the operator* D˜D∼ *can be considered as a unique bounded linear extension of the Dirac operator.*

Lemma 2.4 **(see [28])** ***Let* p(x)?P(Ω)**

(i) *If* $u \in W_0^{1,p(x)}(\Omega, C\ell_n),$ *then the equation* TDu(x)=u(x) *holds for all* x∈Ωx∈Ω.

(ii)If $u \in L^{p(x)}(\Omega, C\ell_n),$*then the equation* DTu(x)=u(x)DTu(x)=u(x) *holds for all* x∈Ωx∈Ω.

Lemma **2.5(see [28])** ***Let*** **p(x)p(x)** ***satisfies*** **(2.1).**

(i)If $u \in L^{p(x)}(\Omega, C\ell_n)$, *then the equation* $\widetilde{T}\widetilde{D}u(x) = u(x)$ *holds for all* x∈Ωx∈Ω.

(ii)If $u \in W^{-1,p(x)}(\Omega, C\ell_n)$, *then the equation* $\widetilde{D}\widetilde{T}u(x) = u(x)$ *holds for all* $x \in \Omega$.

3 A MODIFIED DECOMPOSITION OF SPACES

Gürlebeck and Sprößig [32, 33] showed that the orthogonal decomposition of the space L2(Ω)L2(Ω) holds in the hyper-complex function theory:

$$L^2(\Omega, C\ell_n) = \left(\ker D \cap L^2(\Omega, C\ell_n)\right) \oplus DW_0^{1,2}(\Omega, C\ell_n) \qquad (3.1)$$

with respect to the Clifford-valued product (2.3). Kähler [37] extended the orthogonal decomposition (3.1) to the spaces Lp(Ω,Cℓn) in form of a direct decomposition in a bounded domain. In [38], Cerejeiras and Kähler investigated a direct decomposition of Lp(Ω,Cℓn) in an unbounded domain. In [28], Zhang *et al.* generalized (3.1) to the setting of Lp(x)(Ω,Cℓn) in a bounded domain.

In [36], Diening *et al.* showed that the Dirichlet problem of the Poisson equation with homogeneous boundary data

$$\begin{cases} -\Delta u = f & \text{in } \Omega, \\ u = 0 & \text{on } \partial\Omega, \end{cases} \qquad (3.2)$$

has a unique weak solution u∈W1,p(x)(Ω)u∈W1,p(x)(Ω) for each f∈W−1,p(x)(Ω)f∈W−1,p(x)(Ω). Moreover, the following estimate holds:

$$\|u\|_{W^{1,p(x)}(\Omega)} \leq C(n,p,\Omega)\|f\|_{W^{-1,p(x)}(\Omega)}.$$

Here *u* is called a *weak solution* of problem (3.2) provided that

$$\langle f,\varphi\rangle = \int_\Omega \nabla u\cdot\nabla\varphi\,dx, \quad \forall\varphi\in W_0^{1,p'(x)}(\Omega).$$

Then it is easy to see that for all $f\in W^{-1,p(x)}(\Omega, C\ell_n)$ the problem (3.2) still has a unique weak solution $u\in W^{1,p(x)}(\Omega, C\ell_n)$. We denote by Δ−10Δ0−1 the solution operator.

Now we are in a position to prove a decomposition of the variable exponent Lebesgue spaces equipped with the modified inner product.

Theorem 3.1

The space $L^{p(x)}(\Omega, C\ell_n)$ *allows the Hodge-type decomposition*

$$L^{p(x)}(\Omega, C\ell_n) = \left(B\ker\widetilde{D}\cap L^{p(x)}(\Omega, C\ell_n)\right)\oplus DW_0^{1,p(x)}(\Omega, C\ell_n) \tag{3.3}$$

with respect to the inner product

$$(f,g)_A = \int_\Omega \overline{Af(x)}Ag(x)\,dx,$$

where AB=BA=I *and* Bu:=b(x,u)u *with* ab=1.

Proof

Similar to the proof of Theorem 6 in [37], we first show the intersection of spaces $B\ker\widetilde{D}\cap L^{p(x)}(\Omega, C\ell_n)$ and spaces $DW_0^{1,p(x)}(\Omega, C\ell_n)$ and spaces $\cdot\widetilde{D}\cap L^{p(x)}(\Omega, C\ell_n)$ and spaces $DW_0^{1,p(x)}(\Omega, C\ell_n)$contains one and only one zero element. Indeed, suppose $f\in(B\ker\widetilde{D}\cap L^{p(x)}(\Omega, C\ell_n))\cap DW_0^{1,p(x)}(\Omega, C\ell_n)$, then $\widetilde{D}Af = 0$. Moreover,

$f \in L^{p(x)}(\Omega, C\ell_n)$ shows that there exists a function $w = TAf \in W^{1,p(x)}(\Omega, C\ell_n)$ because of Lemma 2.2; then $f = BDw \in DW_0^{1,p(x)}(\Omega, C\ell_n)$ due to Lemma 2.4. It follows that w∈W1,p(x)0(Ω,Cℓn)w∈W01,p(x)(Ω,Cℓn). Therefore, we have D˜Dw=0D~Dw=0 in Ω and w=0w=0 on ∂Ω. From the uniqueness of Δ_0^{-1} we obtain $w = 0$.. Consequently, f=0f=0. Therefore, the sum of the two subspaces is a direct one.

Now let $u \in L^{p(x)}(\Omega, C\ell_n)$. If we take $u_2 = BD\Delta_0^{-1}\widetilde{D}Au$, then using the same arguments as the first part, we deduce that $u_k \in W_0^{1,p(x)}(\Omega, C\ell_n)$ such that $u_k \to u$ in $L^{p(x)}(\Omega, C\ell_n)$,. Then u1∈Lp(x)(Ω,Cℓn)u1∈Lp(x)(Ω,Cℓn). Furthermore, we take $u_k \to u$ in $L^{p(x)}(\Omega, C\ell_n)$, such that uk→uuk→u in Lp(x)(Ω,Cℓn)Lp(x)(Ω,Cℓn), then by the density of $W_0^{1,p(x)}(\Omega, C\ell_n)$ in $L^{p(x)}(\Omega, C\ell_n)$ and Lemma 2.3, we have for any φ∈W1,p′(x)0(Ω,Cℓn)φ∈W01,p′(x)(Ω,Cℓn),

$$(u_1, BD\varphi)_A = \int_\Omega \overline{A(u - u_2)D\varphi}\, dx = \lim_{k\to\infty} \int_\Omega \overline{(DAu_k - DABD\Delta_0^{-1}DAu_k)\varphi}\, dx =$$

Thus, we obtain u1∈BkerD˜u1∈Bker D~. Since u∈Lp(x)(Ω,Cℓn) u∈Lp(x)(Ω,Cℓn) is arbitrary, the desired result follows. □

From this decomposition we can get the following projections:

$$P_a : L^{p(x)}(\Omega, C\ell_n) \to B\ker\widetilde{D} \cap L^{p(x)}(\Omega, C\ell_n),$$

$$Q_a : L^{p(x)}(\Omega, C\ell_n) \to DW_0^{1,p(x)}(\Omega, C\ell_n).$$

Moreover, we have

$$Q_a = BD\Delta_0^{-1}\widetilde{D}A, \qquad P_a = I - Q_a. \tag{3.4}$$

Corollary 3.1

$f \in L^{p(x)}(\Omega, C\ell_n)$.
. *Then the following equations with homogeneous boundary data*:

$$\begin{cases} DADu = f & in\ \Omega, \\ u = 0 & on\ \partial\Omega \end{cases} \tag{3.5}$$

has a unique solution u∈W2,p(x)0(Ω,Cℓn)u∈W02,p(x)(Ω,Cℓn) *which may be represented by the formula* u=TQaBTfu=TQaBTf.

Proof

Proof For existence of solution, Theorem 3.1 implies that there exists a function $u \in W_0^{1,p(x)}(\Omega, C\ell_n)$ such that $Q_aBTf = Du$. Lemma 2.2 gives $Tf \in W^{1,p(x)}(\Omega, C\ell_n)$ and $Q_aBTf \in W_0^{2,p(x)}(\Omega, C\ell_n)$. Then from Lemma 2.4 it follows that $TQ_aBTf = TDu = u$. Hence $u \in W_0^{2,p(x)}(\Omega, C\ell_n)$. Further, we have

$$DADu = DAQ_aBTf = DABTf - DAP_aBf = f,$$

which implies DADu=f

For the uniqueness of solution, consider the boundary value problems: DADu=0 in Ω and u=0 on ∂Ω. It is easy to see that $Du \in \operatorname{im} P_a \cap \operatorname{im} Q_a = \{0\}$. Then $Du = 0$. because of Lemma 2.4. □

Remark 3.1

Gürlebeck and Sprößig [33] pointed out that the equations of linear elasticity and time-independent Maxwell equations over a three-dimensional domain with zero boundary condition could be transformed into problem (3.5) in a quaternionic language; see Section 4.3 and Section 4.4 in [33] for more details.

Corollary 3.2

The space $L^{p(x)}(\Omega, C\ell_n) \cap \operatorname{im} Q_a$ im□Qa *is a closed subspace of* $L^{p(x)}(\Omega, C\ell_n)$.

Proof

The proof follows that of Lemma 6 in [28] almost word by word, we thus omit the details. □

Corollary 3.3

$(L^{p(x)}(\Omega, C\ell_n) \cap \operatorname{im} Q_a)^* = L^{p'(x)}(\Omega, C\ell_n) \cap \operatorname{im} Q_a$. *In other words, the linear operator*

$$\Phi : DW_0^{1,p'(x)}(\Omega, C\ell_n) \to \left(DW_0^{1,p(x)}(\Omega, C\ell_n)\right)^*$$

given by

$$\Phi(Du)(D\varphi) = [D\varphi, Du]_A := \int_\Omega [\overline{AD\varphi}ADu]_0 \, dx$$

is a Banach space isomorphism.

Proof

Similar to the proof of Lemma 2.7 in [28], for the reader's convenience, we will give a detailed treatment of the proof. In view of Corollary 3.2, $DW_0^{1,p(x)}(\Omega, C\ell_n)$ and $DW_0^{1,p(x)}(\Omega, C\ell_n)$ are reflexive Banach spaces since they are closed in $L^{p(x)}(\Omega, C\ell_n)$ and $L^{p'(x)}(\Omega, C\ell_n)$,, respectively. Linearity of Φ is clear. For injectivity, suppose

$$\Phi(Du)(D\varphi) = [D\varphi, Du]_A = 0 \tag{3.6}$$

for all $\varphi \in W_0^{1,p(x)}(\Omega, C\ell_n)$ and some $u \in W_0^{1,p'(x)}(\Omega, C\ell_n)$. For any $\omega \in L^{p(x)}(\Omega, C\ell_n)$, ac-, according to (3.4), we may write ω=α+βω=α+β with αϵBkerD˜∩Lp(x)(Ω,Cℓn)αϵBker $\widetilde{D} \cap L^{p(x)}(\Omega, C\ell_n)$ and βϵDW1,p(x)0(Ω,Cℓn)βϵDW01,p(x)(Ω,Cℓn). Thus we obtain

$$[\omega, Du]_A = [\alpha + \beta, Du]_A = [\alpha, Du]_A + [\beta, Du]_A.$$

Taking a similar argument to the first part of the proof in Theorem 3.1, we may write α=B $\zeta \in \ker \widetilde{D} \cap L^{p(x)}(\Omega, C\ell_n)$and $Du = BD\eta$ with $\eta \in W_0^{1,p(x)}(\Omega, C\ell_n)$, then it is easy to see that (α,Du)Sc=00. This together with (3.6) gives (ω,Du)Sc=0. This leads to Du=0. It follows that Φ is injective. To get surjectivity, let $f \in (DW_0^{1,p(x)}(\Omega, C\ell_n))^*$. By the Hahn-Banach theorem, there is $F \in (L^{p(x)}(\Omega, C\ell_n))^*$ with $\|F\| = \|f\|$ and $F|_{DW_0^{1,p(x)}(\Omega, C\ell_n)} = f$. Moreover, there exists $\varphi \in L^{p'(x)}(\Omega, C\ell_n)$ such that $F(u) = (u, \varphi)_{Sc}$ for any $u \in L^{p(x)}(\Omega, C\ell_n)$. According to (3.4), we can write $\varphi = \xi + D\alpha$, where $\xi \in B\ker \widetilde{D} \cap L^{p'(x)}(\Omega, C\ell_n)$, $D\alpha \in DW_0^{1,p'(x)}(\Omega, C\ell_n)$. For any $Du \in DW_0^{1,p(x)}(\Omega, C\ell_n)$, we have

$$f(Du) = [Du, \varphi]_A = [Du, \xi]_A + [Du, D\alpha]_A = [Du, D\alpha]_A = \Phi(D\alpha)(Du).$$

Consequently, Φ(Dα)=fΦ(Dα)=f. It follows that Φ is surjective. By Theorem 3.1 in [23] we have

$$\left|\Phi(Du)(D\varphi)\right| = \left|[D\varphi, Du]_A\right| \le C\|D\varphi\|_{L^{p(x)}(\Omega, C\ell_n)}\|Du\|_{L^{p'(x)}(\Omega, C\ell_n)}.$$

This means that Φ is continuous. Furthermore, it is immediate that Φ−1Φ−1 is continuous from the inverse function theorem. The proof of Corollary 3.3 is thus finished. □

4 STOKES EQUATIONS WITH VARIABLE VISCOSITY

In the study of the stationary Navier-Stokes equations, the corresponding Stokes equations play an important role. It can be said that any open question about the Navier-Stokes equations, such as global existence of strong solutions, uniqueness and regularity of weak solutions, and asymptotic behavior, is closely related with the qualitative and quantitative properties of the solutions of Stokes equations; see, for example, [38] for related discussions. To be precise, it is crucial to investigate the properties of solutions (u,π)(u,π) to the following Stokes system:

$$-\nu\Delta u + \nabla\pi = f \quad \text{in } \Omega, \tag{4.1}$$

$$\operatorname{div} u = f_0 \quad \text{in } \Omega, \tag{4.2}$$

$$u = v_0 \quad \text{on } \partial\Omega. \tag{4.3}$$

With $\int_\Omega f_0\,dx = \int_{\partial\Omega} n \cdot v_0\,dx$, the necessary condition for the solvability is given. The scalar function f0f0 is a measure of the compressibility of fluid. The boundary condition (4.3) describes the adhesion at the boundary of the domain Ω for v0=0v0=0. This system describes the stationary flow of a homogeneous viscous fluid for small Reynold's numbers. For more details we refer to [3,33].

For $f = \sum_{i=1}^{n} f_i e_i$ and $u = \sum_{i=1}^{n} u_i e_i$, let us consider the following Stokes system in the hyper-complex formulation:

$$\widetilde{D}ADu + D\pi = f \quad \text{in } \Omega, \tag{4.4}$$

$$[Du]_0 = 0 \quad \text{in } \Omega, \tag{4.5}$$

$$u = 0 \quad \text{on } \partial\Omega. \tag{4.6}$$

Definition 4.1 **We say that** $(u,\pi) \in W_0^{1,p(x)}(\Omega, C\ell_n) \times L^{p(x)}(\Omega)$ **a solution of (4.4)-(4.6) provided that it satisfies the system (4.4)-(4.6) for every** $f \in W^{-1,p(x)}(\Omega, C\ell_n)$.

Definition 4.2 **The operator** $\widetilde{\nabla} : L^{p(x)}(\Omega) \to (W^{-1,p(x)}(\Omega))^n$ **n is defined by**

$$\langle \widetilde{\nabla} f, \varphi \rangle = -\langle f, \operatorname{div} \varphi \rangle := -\int_\Omega f \operatorname{div} \varphi \, dx$$

for all $f \in L^{p(x)}(\Omega)$ and $\varphi \in (C_0^\infty(\Omega))^n$.

Lemma 4.1 **(see [28])** ***Let Ω be a bounded Lipschitz domain of*** $\mathbb{R}^n$. *Let* $f \in (W^{-1,p(x)}(\Omega))^n$ ***satisfy***

$$\langle f, \varphi \rangle = \int_\Omega f \cdot \varphi \, dx = 0$$

for any $\varphi \in \mathcal{M}(\Omega) := \{v \in (W_0^{1,p'(x)}(\Omega))^n : \operatorname{div} v = 0\}$. *Then there exists* q∈Lp(x)(Ω) *such that* f=∇˜qf=∇~q.

Theorem 4.1

Let $f \in W^{-1,p(x)}(\Omega, C\ell_n)$. *Then the Stokes system* (4.4)-(4.6) *has a unique solution* $(u,\pi) \in W_0^{1,p(x)}(\Omega, C\ell_n) \times L^{p(x)}(\Omega)$ *in the form*

$$u + TQ_a B\pi = TQ_a B\widetilde{T} f,$$

with respect to the estimate

$$\|Du\|_{L^{p(x)}(\Omega, C\ell_n)} + \|Q_a B\pi\|_{L^{p(x)}(\Omega)} \leq C \|Q_a B\widetilde{T} f\|_{L^{p(x)}(\Omega, C\ell_n)}.$$

Here, C≥1C≥1 *is a constant and the hydrostatic pressure π is unique up to a constant.*

Proof

We first show that the following representation is valid for each $f \in W^{-1,p(x)}(\Omega, C\ell_n)$:

$$TQ_aB\widetilde{T}f = u + TQ_aB\pi .$$

Indeed, let $\varphi_n \in W_0^{1,p(x)}(\Omega, C\ell_n)$ with $\varphi_n \to \varphi$ in $L^{p(x)}(\Omega, C\ell_n)$. By Lemma 2.4, we have

$$TQBT(D\varphi_n) = TQ_aB\varphi_n .$$

Since $W_0^{1,p(x)}(\Omega, C\ell_n)$ is dense in $L^{p(x)}(\Omega, C\ell_n)$, it follows that $TQ_aB\widetilde{T}\widetilde{D}\varphi = TQ_aB\varphi$. Thus, for $u \in W_0^{1,p(x)}(\Omega, C\ell_n)$ and $\pi \in L^{p(x)}(\Omega)$, Lemma 2.4 yields

$$TQ_aB\widetilde{T}f = TQ_aB\widetilde{T}(\widetilde{D}ADu + \widetilde{D}\pi) = u + TQ_aB\pi .$$

This implies that our system (4.4)-(4.6) is equivalent to the system

$$u + TQ_aB\pi = TQ_aB\widetilde{T}f, \tag{4.7}$$

$$[Q_aB\pi]_0 = [Q_aB\widetilde{T}f]_0 . \tag{4.8}$$

Clearly, the equality (4.7) is equivalent to the following equality:

$$Du + Q_aB\pi = Q_aB\widetilde{T}f . \tag{4.9}$$

Now we need to show that for each $f \in W^{-1,p(x)}(\Omega, C\ell_n^1)$, the function QaBTf can be decomposed into two functions *Du*and QaBπQaBπ.

Suppose $Du + Q_a B\pi = 0$ for $u \in$ for $W_0^{1,p(x)}(\Omega, C\ell_n^1) \cap \ker \operatorname{div}$ and $\pi \in L^{p(x)}(\Omega)$. Then (4.5) gives [QaBπ]0=0 . Thus, QaBπ=0 . Hence, Du=QaBπ=0 . This means that Du+QaBπ is a direct sum, which is a subset of imQa.

Next we have to consider the existence of a functional F∈(Lp(x)(Ω,Cℓ1n)∩imQa)∗ with F(Du)=0 and F(QaBπ)=0 but F(QaBT˜f)≠0 . This is equivalent to asking if there exists g∈W−1,p′(x)(Ω,Cℓ1n) such that for all u∈W1,p(x)0(Ω,Cℓ1n)∩kerdiv ker□div and π∈Lp(x)(Ω) ,

$$[Du, Q_a B\widetilde{T}g]_A = 0, \tag{4.10}$$

$$[Q_a B\pi, Q_a B\widetilde{T}g]_A = 0, \tag{4.11}$$

$[Q_a B\widetilde{T}f, Q_a B\widetilde{T}g]_A \neq 0.$. Here, Corollary 3.3 is applied.

Thus, let us consider the system (4.10) and (4.11) with $g \in W^{-1,p'(x)}(\Omega, C\ell_n^1)$. Taking the arguments as the first part of the proof in Theorem 3.1, we may write Du=BDw with $w \in W_0^{1,p(x)}(\Omega, C\ell_n^1)$.. Then divu=divw=0div□u=div□w=0 because of b(x)>0 . According to Lemma 2.5, (4.10) yields

$$\begin{aligned} 0 = [Du, Q_a B\widetilde{T}g]_A &= \int_\Omega \overline{ADu} A Q_a B\widetilde{T}g\,dx \\ &= \int_\Omega \overline{w}\widetilde{D} A Q_a B\widetilde{T}g\,dx = \int_\Omega \overline{w}g\,dx - \int_\Omega \overline{w}\widetilde{D} A P_a B\widetilde{T}g\,dx = \int_\Omega \overline{w}g\,dx, \end{aligned}$$

which implies $g = \widetilde{\nabla}h = \widetilde{D}h$ with $h \in L^{p'(x)}(\Omega)$ because of Lemma 4.1. Thus we obtain from (4.11) and Lemma 2.4

$$[Q_a B\pi, Q_a B\widetilde{T}g]_A = [Q_a B\pi, Q_a B\widetilde{T}\widetilde{D}h]_A = [Q_a B\pi, Q_a Bh]_A = 0$$

holds for each $\pi \in L^{p(x)}(\Omega)$. Hence, $Q_a B\pi = |Q_a Bh|^{p'(x)-2} Q_a Bh$ gives $Q_a Bh = 0$. Then we obtain

$$g = \widetilde{D}h = \widetilde{D}AQ_aBh + \widetilde{D}AP_aBh = 0.$$

Furthermore, we get

$$[Q_aB\widetilde{T}f, Q_aB\widetilde{T}g]_A = 0, \quad \forall f \in W^{-1,p(x)}(\Omega, C\ell_n^1).$$

Finally, (4.9) yields

$$\|Du\|_{L^{p(x)}(\Omega,C\ell_n)} + \|Q_aB\pi\|_{L^{p(x)}(\Omega)} \geq \|Q_aB\widetilde{T}f\|_{L^{p(x)}(\Omega,C\ell_n)}.$$

By the norm equivalence theorem, we obtain

$$\|Du\|_{L^{p(x)}(\Omega,C\ell_n)} + \|Q_aB\pi\|_{L^{p(x)}(\Omega)} \leq C\|Q_aB\widetilde{T}f\|_{L^{p(x)}(\Omega,C\ell_n)}.$$

By Remark 2.1, Lemma 2.2, and the boundedness of the operator QaQa, we get

$$\|u\|_{W_0^{1,p(x)}(\Omega,C\ell_n)} + \|Q_aB\pi\|_{L^{p(x)}(\Omega)} \leq C\|f\|_{W^{-1,p(x)}(\Omega,C\ell_n)}. \tag{4.12}$$

From (4.12) the uniqueness of the solution follows. Note that QaB(π1−π2)=0QaB(π1−π2)=0 implies B(π1−π2)∈BkerD˜B (π1−π2) ∈Bker⁡D~. Then π1=π2+cπ1=π2+c with c∈Rc∈R. Therefore, π is unique up to a constant. The proof of Theorem 4.1 is now complete. □

In [28], the authors did not point out the relation of solutions and weak solutions for the Stokes equations. Next we should prove that solutions and weak solutions for the Stokes equations are equivalent.

Definition 4.3 **The Clifford-valued function *u* is called a weak solution of the system (4.3)-(4.6), if for each** $v \in W_0^{1,p'(x)}(\Omega, C\ell_n^1) \cap$ **ker⍰div, the equality**

$$(ADu, Dv)_{Sc} = (f, v)_{Sc}$$

holds, where (u,v)Sc(u,v)Sc denotes the classical scalar product $(u,v)_{Sc} = \int_\Omega [\overline{u}v]_0\, dx = \int_\Omega \sum_{i=1}^n u_i \times$.

Theorem 4.2 **Let** $u \in W_0^{1,p(x)}(\Omega, C\ell_n^1)$ *and* $\pi \in L^{p(x)}(\Omega)$ ***be a solution of system* (4.7)-(4.8).** ***Then u is a weak solution of system* (4.4)-(4.6).*If u is a weak solution of the system* (4.4)-(4.6), *then there exists a function* π** $\in W_0^{1,p'(x)}(\Omega, C\ell_n^1) \cap$ ***such that the pair* (u,π) *solves the system* (4.7)-(4.8).**

Proof **We follow the idea of Theorem 4.5.2 in [33]. First we need to show that for all functions v⍰W1,p′ (x)0 (Ω,Cℓ1n)∩ kerdivv⍰W01,p′(x)(Ω,Cℓn1)∩ker⍰div, the following equality holds:**

$$\sum_{i=1}^{n} (A\nabla u_i, \nabla v_i)_{Sc} = (f, v)_{Sc}. \tag{4.13}$$

For this purpose, from (4.7) it follows that

$$(ADu, Dv)_{Sc} = (ADTQ_a B\widetilde{T} f, Dv)_{Sc} - (ADTQ_a B\pi, Dv)_{Sc} \tag{4.14}$$

for each $v \in W_0^{1,p'(x)}(\Omega, C\ell_n^1) \cap$ ker⍰div. The first term of the right side in (4.14) yields

$$(ADTQ_a B\widetilde{T} f, Dv)_{Sc} = (\widetilde{D} ADTQ_a B\widetilde{T} f, v)_{Sc} = (f, v)_{Sc}, \tag{4.15}$$

thanks to Lemma 2.4, Lemma 2.5 and (3.4). For the second term of the right side in (4.14), we choose a sequence $\{\pi_k\}_{k=1}^{\infty} \subset W_0^{1,p(x)}(\Omega)$ with $\pi_k \to \pi$ in $L^{p(x)}(\Omega)$, then

$$\sum_{i=1}^{n}\big((ADTQ_aB\pi)_i, Dv_i\big)_{Sc} = \lim_{k\to\infty}\sum_{i=1}^{n}\big((ADTQ_aB\pi_k)_i, Dv_i\big)_{Sc}$$

$$= \lim_{k\to\infty}(DADTQ_aB\pi_k, v)_{Sc} = \lim_{k\to\infty}(D\pi_k, v)_{Sc} = 0, \tag{4.16}$$

because $(\nabla\pi_k, v)_{Sc} = -(\pi_k, \operatorname{div} v)_{Sc} = 0$ for all $v \in W_0^{1,p'(x)}(\Omega, C\ell_n^1) \cap$ ker div. Then (4.14), together with (4.15) and (4.16), gives (4.13).

Conversely, let $u \in W_0^{1,p(x)}(\Omega, C\ell_n^1) \cap$ ker div be a weak solution for the system (4.4)-(4.6). That means that (4.13) holds. It easily follows from Lemma 2.4 and Theorem 3.1 that

$$(ADu, Dv)_{Sc} - (ADTQ_aB\widetilde{T}f, Dv)_{Sc} + (ADTQ_aBq, Dv)_{Sc} = 0$$

for all $q \in L^{p(x)}(\Omega)$ and $v \in W_0^{1,p'(x)}(\Omega, C\ell_n^1) \cap$ ker div. Then we have

$$\sum_{i=1}^{n}\big(\nabla(u - TQ_aB\widetilde{T}f + TQ_aBq)_i, A\nabla v_i\big)_{Sc} = 0. \tag{4.17}$$

On one hand, we have shown in Theorem 4.1, which is independent of the above considerations, that the system

$$\psi + TQ_aB\pi = TQ_aB\widetilde{T}f,$$

$$[Q_aB\pi]_0 = [Q_aB\widetilde{T}f]_0$$

has a solution $(\psi, \pi) \in W_0^{1,p(x)}(\Omega, C\ell_n^1) \times L^{p(x)}(\Omega)$. Then substituting q in (4.17) by π, we obtain

$$[Du - Q_a B\widetilde{T}f + Q_a B\pi]_0 = 0,$$

which implies

$$u - TQ_a B\widetilde{T}f + TQ_a B\pi \in W_0^{1,p(x)}\left(\Omega, C\ell_n^1\right) \cap \ker \operatorname{div}.$$

On the other hand, it follows from (4.17) that

$$\left(DAD(u - TQ_a B\widetilde{T}f + TQ_a B\pi), v\right)_{Sc} = 0 \tag{4.18}$$

for all $v \in W_0^{1,p'(x)}(\Omega, C\ell_n^1) \cap \ker \operatorname{div}$. Obviously, $W_0^{1,p'(x)}(\Omega, C\ell_n^1) \cap \ker \operatorname{div}$ is a closed subspace in $W_0^{1,p'(x)}(\Omega, C\ell_n^1)$, and hence $W_0^{1,p'(x)}(\Omega, C\ell_n^1) \cap \ker \operatorname{div}$ is a reflexive Banach space due to Lemma 2.1. Then (4.18) yields

$$DAD(u - TQ_a B\widetilde{T}f + TQ_a B\pi) = 0.$$

Finally, by using the density argument and Corollary 3.1 we obtain

$$u - TQ_a B\widetilde{T}f + TQ_a B\pi = 0.$$

This ends the proof of Theorem 4.2. □

Remark 4.1 **Actually, the last part of the proof in Theorem 4.2 is quite different from that of Theorem 4.5.2 in [33]. The reason lies in the difference of** $W_0^{1,2}(\Omega, C\ell_n)$ **and** $W_0^{1,p(x)}(\Omega, C\ell_n)$.

5 NAVIER-STOKES EQUATIONS WITH VARIABLE VISCOSITY

Our aim in this section is to investigate the existence of solution of the time-independent generalized Navier-Stokes equations (1.7). Evidently, the main difference from the above-mentioned Stokes equations is the appearance of the nonlinear convection term (u·∇)u(u·∇)u. In 1928, Oseen showed that one can get relatively good results if the convection term $(u\cdot\nabla)u(u\cdot\nabla)u$ is replaced by $(v\cdot\nabla)u(v\cdot\nabla)u$, where v is a solution of the corresponding Stokes equations. In 1965, Finn [39] proved the existence of solutions for small external forces with a spatial decreasing to infinity of order $|x|-1|x|-1$ for the case of $n=3n=3$ and used the Banach fixed-pointed theorem. Gürlebeck and Sprößig [32, 33] solved this system by a reduction to a sequence of Stokes problems provided the external force f belongs to $Lp(\Omega,H)Lp(\Omega,H)$ for a bounded domain Ω and $6/5<p<3/26/5<p<3/2$. Cerejeiras and Kähler [38] obtained similar results provided the external force f belongs to $W-1,p(\Omega,C\ell n)W-1,p(\Omega,C\ell n)$ for an unbounded domain Ω and $n/2\leq p<\infty n/2\leq p<\infty$. Zhang *et al.*[28] investigated similar results provided that the external force f belongs to $W-1,p(x)(\Omega,C\ell n)W-1,p(x)(\Omega,C\ell n)$ for a bounded domain Ω and $n/2\leq p<\infty n/2\leq p<\infty$. Our intention in this section is to extend these results to the more general case in variable exponent spaces.

For $f=\sum ni=1fieif=\sum i=1nfiei$, $u=\sum ni=1uieiu=\sum i=1nuiei$, let us consider the following steady generalized Navier-Stokes equations in the hyper-complex notation:

$$\widetilde{D}ADu + D\pi = F(u) \quad \text{in } \Omega, \tag{5.1}$$

$$[Du]_0 = 0 \quad \text{in } \Omega, \tag{5.2}$$

$$u = 0 \quad \text{on } \partial\Omega, \tag{5.3}$$

with the nonlinear part $F(u)=f-[uD]0u$, where the operator A is mentioned above.

Lemma 5.1 **(see [28])** ***Let*** **p(x)p(x)** ***satisfies*** **(2.1)** ***and*** **n/2≤p−≤p(x)≤p+<∞** ***Then the operator*** $F: W_0^{1,p(x)}(\Omega, C\ell_n^1) \to W^{-1,p(x)}(\Omega, C\ell_n^1)$ **Cℓn1)** ***is a continuous operator and we have***

$$\left\| [uD]_0 u \right\|_{W^{-1,p(x)}(\Omega, C\ell_n)} \leq C_1 \|u\|^2_{W_0^{1,p(x)}(\Omega, C\ell_n)},$$

where C1=C1(n,p,Ω)C1=C1(n,p,Ω) *is a positive constant.*

Using Lemma 5.1, which is crucial to the convergence of the iteration method, we are able to give the main result as follows.

Theorem 5.1

Let p(x)p(x) *satisfy* (2.1) *and* n/2≤p−≤p(x)≤p+<∞. *Then the system* (5.1)-(5.3) *has a unique solution* ($(u, \pi) \in W_0^{1,p(x)}(\Omega, C\ell_n) \times L^{p(x)}(\Omega, \mathbb{R})$ (*π is unique up to a real constant*) *if the right-hand side f satisfies the condition*

$$\|f\|_{W^{-1,p(x)}(\Omega, C\ell_n)} < \frac{\nu^2}{4C_1C_4^2}, \tag{5.4}$$

with ν=μ/ρν=μ/ρ, C4=C2(1+C3)C4=C2(1+C3), *where* C3≥1C3≥1 *indicated in* (5.8) *and*

$$C_2 = \|T\|_{[L^{p(x)} \cap \operatorname{im} Q, W_0^{1,p(x)}]} \|QB\|_{[L^{p(x)}, L^{p(x)} \cap \operatorname{im} Q]} \|\widetilde{T}\|_{[W^{-1,p(x)}, L^{p(x)} \cap \operatorname{im} Q]}.$$

For any function $u_0 \in W_0^{1,p(x)}(\Omega, C\ell_n)$ *with*

$$\|u_0\|_{W_0^{1,p(x)}(\Omega, C\ell_n)} \leq \frac{\nu}{2C_1C_2} - \mathcal{F}, \tag{5.5}$$

here $\mathcal{F} = \sqrt{\frac{\nu^2}{4C_1^2C_4^2} - \frac{1}{C_1}\|f\|_{W^{-1,p(x)}(\Omega, C\ell_n)}},$ *the iteration procedure*

$$u_k + TQB_{k-1}\pi_k = TQB_{k-1}\widetilde{T}F(u_{k-1}), \quad k = 1, 2, \ldots,$$
$$[QB_{k-1}\pi_k]_0 = \left[QB_{k-1}\widetilde{T}F(u_{k-1})\right]_0, \tag{5.6}$$

$$[QB_{k-1}\pi_k]_0 = \left[QB_{k-1}\widetilde{T}F(u_{k-1})\right]_0, \tag{5.7}$$

converges in $W_0^{1,p(x)}(\Omega, C\ell_n) \times L^{p(x)}(\Omega)$, *where* $B_k h := b(x, u_k)h$.

Proof **The proof is similar to that of Theorem 4.6.8 in [33]. For the reader's convenience, we will give the key details of the proof. Replacing *f* by F(uk−1)F(uk−1) in the proof of Theorem 4.1, we obtain the unique solvability of the Stokes equations (4.4)-(4.6) which we have to solve in each step. Moreover, we have the following estimate:**

$$\|Du_k\|_{L^{p(x)}(\Omega, C\ell_n)} + \|QB\pi_k\|_{L^{p(x)}(\Omega)} \le C_3 \left\|QB\widetilde{T}F(u_{k-1})\right\|_{L^{p(x)}(\Omega, C\ell_n)}, \tag{5.8}$$

where C3≥1 is a constant. In the following, using the Banach fixed point theorem, we could finish the remaining proof by following the proof of Theorem 3.1 in [28] word by word; we thus omit the details.

□

Definition 5.1 **The Clifford-valued function *u* is called a weak solution of the system (5.6)-(5.7), if for all** $v \in W_0^{1,p(x)}(\Omega, C\ell_n^1) \cap$ **ker⁡div, the following equality holds:**

$$(ADu, Dv)_{Sc} = \left(F(u), v\right)_{Sc},$$

that is to say, the following equality is valid:

$$\sum_{i=1}^{n} (A\nabla u_i, \nabla v_i)_{Sc} + \sum_{i=1}^{n} (u_i \partial_i u, v)_{Sc} = (f, v)_{Sc}$$

for every $v \in W_0^{1,p(x)}(\Omega, C\ell_n^1) \cap$ ker⁡div.

In the following we would like to point out the equivalence of solutions and weak solutions for the generalized Navier-Stokes equations; this point is similar to the case of the Stokes equations.

Theorem 5.2 **Let** $u \in W_0^{1,p(x)}(\Omega, C\ell_n^1)$ *and* $\pi \in L^{p(x)}(\Omega)$ ***be a solution of system* (5.6)-(5.7). *Then u is a weak solution of system* (5.1)-(5.3).*If u is a weak solution of the system* (5.6)-(5.7), *then there exists a function* π⍰Lp(x)(Ω)π⍰Lp(x)(Ω) *such that the pair* (u,π)(u,π) *solves the system* (5.1)-(5.3).**

Proof **The proof is quite similar to the proof of Theorem 4.2, so we omit it. We also refer the reader to a similar proof of Theorem 4.6.2 in [33] by replacing *Q* with QaBQaB.** □

ACKNOWLEDGEMENTS

The authors would like to thank the referee for constructive suggestion which helps us to improve the quality of the paper. B Zhang was supported by the Research Foundation of Heilongjiang Educational Committee (Grant No. 12541667).

COMPETING INTERESTS

The authors declare that they have no competing interests.

AUTHORS' CONTRIBUTIONS

All authors carried out the theoretical studies, participated in the design of the study and drafted the manuscript. All authors read and approved the final manuscript.

REFERENCES

1. Acerbi, E, Mingione, G: Regularity results for stationary electro-rheological fluids. Arch. Ration. Mech. Anal. **164**, 213-259 (2002)
2. Sprößig, W: Fluid flow equations with variable viscosity in quaternionic setting. Adv. Appl. Clifford Algebras **17**, 259-272 (2007)
3. Galdi, GP: An Introduction to the Mathematical Theory of the Navier-Stokes Equations. Springer, New York (2011)
4. Rajagopal, KR, Růžička, M: On the modeling of electrorheological materials. Mech. Res. Commun. **23**, 401-407 (1996)
5. Růžička, M: Electrorheological Fluids: Modeling and Mathematical Theory. Springer, Berlin (2000)
6. Gilbert, J, Murray, MAM: Clifford Algebra and Dirac Operators in Harmonic Analysis. Oxford University Press, Oxford (1993)
7. Delanghe, R, Sommen, F, Soucek, V: Clifford Algebra and Spinor-Valued Function. Kluwer Academic, Dordrecht (1992)
8. Doran, C, Lasenby, A: Geometric Algebra for Physicists. Cambridge University Press, Cambridge (2003)
9. Diening, L, Harjulehto, P, Hästö, P, Růžička, M: Lebesgue and Sobolev Spaces with Variable Exponents. Springer, Berlin (2011)
10. Fu, YQ: Weak solution for obstacle problem with variable growth. Nonlinear Anal. **59**, 371-383 (2004)
11. Fu, YQ: The principle of concentration compactness in spaces Lp(x)(Ω)Lp(x)(Ω) and its application. Nonlinear Anal. **71**, 1876-1892 (2009)
12. Rădulescu, V, Zhang, BL: Morse theory and local linking for a nonlinear degenerate problem arising in the theory of

electrorheological fluids. Nonlinear Anal., Real World Appl. **17**, 311-321 (2014)

13. Zhikov, VV: Averaging of functionals of the calculus variations and elasticity theory. Izv. Akad. Nauk SSSR, Ser. Mat. **877**, 675-710 (1986)

14. Zhikov, VV: Solvability of the three-dimensional thermistor problem. Tr. Mat. Inst. Steklova **261**, 101-114 (2008)

15. Kováčik, O, Rákosník, J: On spaces Lp(x)Lp(x) and Wm,p(x)Wm,p(x). Czechoslov. Math. J. **41**, 592-618 (1991)

16. Edmunds, DE, Rákosník, J: Sobolev embedding with variable exponent. Stud. Math. **143**, 267-293 (2000)

17. Edmunds, DE, Rákosník, J: Sobolev embedding with variable exponent II. Math. Nachr. **246**, 53-67 (2002)

18. Fan, XL, Zhao, D: On the spaces Lp(x)Lp(x) and Wm,p(x)Wm,p(x). J. Math. Anal. Appl. **263**, 424-446 (2001)

19. Fan, XL, Shen, JS, Zhao, D: Sobolev embedding theorems for spaces Wk,p(x)(Ω)Wk,p(x)(Ω). J. Math. Anal. Appl. **262**, 749-760 (2001)

20. Harjulehto, P, Hästö, P, Lê, UV, Nuortio, M: Overview of differential equations with non-standard growth. Nonlinear Anal. **72**, 4551-4574 (2010)

21. Nolder, CA: *A*-Harmonic equations and the Dirac operator. J. Inequal. Appl. **2010**, Article ID 124018 (2010)

22. Nolder, CA: Nonlinear *A*-Dirac equations. Adv. Appl. Clifford Algebras **21**, 429-440 (2011)

23. Fu, YQ, Zhang, BL: Clifford valued weighted variable exponent spaces with an application to obstacle problems. Adv. Appl. Clifford Algebras **23**, 363-376 (2013)

24. Fu, YQ, Zhang, BL: Weak solutions for elliptic systems with variable growth in Clifford analysis. Czechoslov. Math. J. **63**, 643-670 (2013)

25. Zhang, BL, Fu, YQ: Weak solutions for *A*-Dirac equations with variable growth in Clifford analysis. Electron. J. Differ. Equ. **2012**, 227 (2012)

26. Lu, Y, Bao, G: The existence of weak solutions to non-homogeneous *A*-Dirac equations with Dirichlet boundary data. Adv. Appl. Clifford Algebras **24**, 151-162 (2014)

27. Fu, YQ, Rădulescu, V, Zhang, BL: Hodge decomposition of variable exponent spaces of Clifford-valued functions and applications to Dirac and Stokes equations. Preprint

28. Zhang, BL, Fu, YQ, Rădulescu, V: The stationary Navier-Stokes equations in variable exponent spaces of Clifford-valued functions. Adv. Appl. Clifford Algebras **24**, 231-252 (2014)

29. Molica Bisci, G, Rădulescu, V, Zhang, BL: Existence of stationary states for *A*-Dirac equations with variable growth. Adv. Appl. Clifford Algebras (2014).

30. Binlin, Z, Ferrara, M, Molica Bisci, G: An overview for nonlinear *A*-Dirac equations with variable growth. BSG Proc. **21**, 14-26 (2014)

31. Diening, L, Málek, J, Steinhauer, M: On Lipschitz truncations of Sobolev function (with variable exponent) and their selected applications. ESAIM Control Optim. Calc. Var. **14**, 211-232 (2008)

32. Gürlebeck, K, Sprößig, W: Quaternionic and Clifford Calculus for Physicists and Engineers. Wiley, New York (1997)

33. Gürlebeck, K, Sprößig, W: Quaternionic Analysis and Elliptic Boundary Value Problems. Birkhäuser, Boston (1990)

34. Ablamowicz, R (ed.): Clifford Algebras and Their Applications in Mathematical Physics. Volume 1: Algebra and Physics. Birkhäuser, Boston (2000)

35. Ryan, J, Sprößig, W (eds.): Clifford Algebras and Their Applications in Mathematical Physics. Volume 2: Clifford Analysis. Birkhäuser, Boston (2000)

36. Diening, L, Lengeler, D, Růžička, M: The Stokes and Poisson problem in variable exponent spaces. Complex Var. Elliptic Equ. **56**, 789-811 (2011)

37. Kähler, U: On a direct decomposition in the space Lp(Ω)Lp(Ω). Z. Anal. Anwend. **4**, 839-848 (1999)

38. Cerejeiras, P, Kähler, U: Elliptic boundary value problems of fluid dynamics over unbounded domains. Math. Methods Appl. Sci. **23**, 81-101 (2000)

39. Finn, R: On the exterior stationary problem for the Navier-Stokes equations and associated perturbation problems. Arch. Ration. Mech. Anal. **19**, 363-406 (1965)

6

INEQUALITIES FOR THE CASORATI CURVATURES OF SLANT SUBMANIFOLDS IN QUATERNIONIC SPACE FORMS

Vladimir Slesar[1] , Bayram ,Sahin[2] and Gabriel-Eduard Vîlcu[3,4]

[1]Department of Mathematics, University of Craiova, Str. Al.I. Cuza, Nr. 13, Craiova, 200585, Romania.

[2]Department of Mathematics, Inonu University, Malatya, 44280, Turkey.

[3]Department of Mathematical Modelling, Economic Analysis and Statistics, Petroleum-Gas University of Ploie,sti, Bulevardul Bucure,sti, Nr. 39, Ploie,sti, 100680, Romania.

[4]Faculty of Mathematics and Computer Science, Research Center in Geometry, Topology and Algebra, University of Bucharest, Str. Academiei, Nr. 14, Sector 1, Bucure,sti, 70109, Romania.

ABSTRACT

In this paper we prove two sharp inequalities that relate the normalized scalar curvature with the Casorati curvature for a slant submanifold in a quaternionic space form. Moreover, we show that in both cases, the equality at all points characterizes the invariantly quasi-umbilical submanifolds.

KEYWORDS

scalar curvature, mean curvature, Casorati curvature shape, operator quaternionic space, form slant submanifold, optimal inequality

1 INTRODUCTION

In a seminal paper published in the early 1990s, Chen [1] established a sharp inequality for a submanifold in a real space form using the scalar curvature and the sectional curvature, both being intrinsic invariants, and squared mean curvature, the main extrinsic invariant, initiating the theory of δ-invariants or the so-called Chen invariants; this turned out to be one of the most interesting modern research topic in differential geometry of submanifolds. These inequalities were further extended to many classes of submanifolds in different ambient spaces (for an extensive and comprehensive survey on this topic see [2]). For example, in a quaternionic Kähler setting, Chen-like inequalities were proved in [3–11] and a set of open problems in the field was proposed recently in [12]. Moreover, new optimal inequalities involving δ-invariants were recently proved in [13–20]. We also note that some interesting inequalities for the length of the second fundamental form of the warped product submanifolds were obtained recently in [21–25].

On the other hand, the Casorati curvature of a submanifold in a Riemannian manifold is an extrinsic invariant defined as the normalized square of the length of the second fundamental form. It is well known that this notion extends the concept of the principal direction of a hypersurface of a Riemannian manifold to submanifolds of a Riemannian manifold and it was preferred by Casorati over the traditional Gauss curvature because corresponds better with the common intuition of curvature [26] (see also [27–30] for the geometrical meaning and the importance of the Casorati curvatures). Therefore it is of great interest to obtain optimal inequalities for the Casorati curvatures of submanifolds in different ambient spaces. We note that in [31], Decu, Haesen and Verstraelen obtained some optimal inequalities involving the scalar curvature and the Casorati curvature of a Riemannian submanifold in a real space form and the holomorphic sectional curvature and the Casorati curvature of a Kähler hypersurface in a complex space form. Moreover, the same authors proved in [32] an inequality in which the scalar curvature is estimated from above by the normalized Casorati curvatures, while Ghişoiu obtained in [33] some inequalities for the Casorati curvatures of slant submanifolds in complex space forms.

In this paper we generalize these inequalities in a quaternionic setting, proving the following result which also solves the Problem 6.7 from [12].

Theorem 1.1 *Let* Mn *be a θ-slant proper submanifold of a quaternionic space form* M¯¯¯¯4m(c). *Then*:

(i)The normalized δ-Casorati curvature δc(n−1) *satisfies*

$$\rho \leq \delta_c(n-1) + \frac{c}{4}\left(1 + \frac{9}{n-1}\cos^2\theta\right). \tag{1}$$

Moreover, the equality sign holds if and only if Mn *is an invariantly quasi-umbilical submanifold with trivial normal connection in* M¯¯¯¯4m(c), *such that with respect to suitable orthonormal tangent frame* {ξ1,...,ξn} *and normal orthonormal frame* {ξn+1,...,ξ4m}, *the shape operators* Ar≡Aξr, r∈{n+1,...,4m}, *take the following forms*:

$$A_{n+1} = \begin{pmatrix} a & 0 & 0 & \cdots & 0 & 0 \\ 0 & a & 0 & \cdots & 0 & 0 \\ 0 & 0 & a & \cdots & 0 & 0 \\ \vdots & \vdots & \vdots & \ddots & \vdots & \vdots \\ 0 & 0 & 0 & \cdots & a & 0 \\ 0 & 0 & 0 & \cdots & 0 & 2a \end{pmatrix}, \qquad A_{n+2} = \cdots = A_{4m} = 0. \tag{2}$$

(ii)The normalized δ-Casorati curvature δ^c(n−1) *satisfies*

$$\rho \leq \hat{\delta}_c(n-1) + \frac{c}{4}\left(1 + \frac{9}{n-1}\cos^2\theta\right). \tag{3}$$

Moreover, the equality sign holds if and only if Mn *is an invariantly quasi-umbilical submanifold with trivial normal connection in* M‾‾‾‾4m(c), *such that with respect to suitable orthonormal tangent frame* {ξ1,...,ξn} *and normal orthonormal frame* {ξn+1,...,ξ4m}, *the shape operators* Ar≡Aξr, r∈{n+1,...,4m}, *take the following forms*:

$$A_{n+1} = \begin{pmatrix} 2a & 0 & 0 & \cdots & 0 & 0 \\ 0 & 2a & 0 & \cdots & 0 & 0 \\ 0 & 0 & 2a & \cdots & 0 & 0 \\ \vdots & \vdots & \vdots & \ddots & \vdots & \vdots \\ 0 & 0 & 0 & \cdots & 2a & 0 \\ 0 & 0 & 0 & \cdots & 0 & a \end{pmatrix}, \qquad A_{n+2} = \cdots = A_{4m} = 0. \tag{4}$$

2 PRELIMINARIES

2.1 Riemannian invariants

In this subsection we recall some basic concepts in Riemannian geometry, using mainly [34].

Let (M‾,g‾) be an *m*-dimensional Riemannian manifold. For an *n*-dimensional Riemannian submanifold *M* of (M‾,g‾), we denote by *g* the metric tensor induced on *M*. If ∇‾ is the Levi-Civita connection on M‾ and ∇ is the covariant differentiation induced on *M*, then the Gauss and Weingarten formulas are given by

$$\overline{\nabla}_X Y = \nabla_X Y + h(X,Y), \quad \forall X, Y \in \Gamma(TM)$$

and

$$\overline{\nabla}_X N = -A_N X + \nabla_X^{\perp} N, \quad \forall X \in \Gamma(TM), \forall N \in \Gamma\left(TM^{\perp}\right),$$

where h is the second fundamental form of M, $\nabla\perp$ is the connection on the normal bundle and AN is the shape operator of $M$with respect to N. The shape operator AN is related to h by

$$g(A_N X, Y) = \overline{g}\left(h(X, Y), N\right)$$

for all X,Y∈Γ(TM) and N∈Γ(TM⊥).

If we denote by R‾‾‾ and R the curvature tensor fields of ∇‾‾‾ and ∇, then we have the Gauss equation:

$$\overline{R}(X, Y, Z, W) = R(X, Y, Z, W) + \overline{g}\left(h(X, W), h(Y, Z)\right) - \overline{g}\left(h(X, Z), h(Y, W)\right) \tag{5}$$

for all X,Y,Z,W∈Γ(TM).

We denote by K(π) the sectional curvature of M associated with a plane section π⊂TpM, p∈M. If {e1,...,en} is an orthonormal basis of the tangent space TpM and {en+1,...,em} is an orthonormal basis of the normal space T⊥pM, then the scalar curvature τ at p is given by

$$\tau(p) = \sum_{1 \leq i < j \leq n} K(e_i \wedge e_j)$$

and the normalized scalar curvature ρ of M is defined as

$$\rho = \frac{2\tau}{n(n-1)}.$$

We denote by H the mean curvature vector, that is,

$$H(p) = \frac{1}{n}\sum_{i=1}^{n} h(e_i, e_i)$$

and we also set

$$h_{ij}^{\alpha} = g\big(h(e_i, e_j), e_\alpha\big), \quad i,j \in \{1,\dots,n\}, \alpha \in \{n+1,\dots,m\}.$$

Then the squared mean curvature of the submanifold *M* in M¯¯¯ is defined by

$$\|H\|^2 = \frac{1}{n^2}\sum_{\alpha=n+1}^{m}\left(\sum_{i=1}^{n} h_{ii}^{\alpha}\right)^2$$

and the squared norm of *h* over dimension *n* is denoted by $\mathcal{C}$ and is called the Casorati curvature of the submanifold *M*. Therefore we have

$$\mathcal{C} = \frac{1}{n}\sum_{\alpha=n+1}^{m}\sum_{i,j=1}^{n}\left(h_{ij}^{\alpha}\right)^2.$$

The submanifold *M* is called totally geodesic if the second fundamental form vanishes identically and totally umbilical if there is a real number λ such that h(X,Y)=λg(X,Y)H for any tangent vectors *X*, *Y* on *M*. If H=0, then the submanifold *M* is said to be minimal.

The submanifold *M* is called invariantly quasi-umbilical if there exist m−n mutually orthogonal unit normal vectors ξn+1,...,ξmsuch that the shape operators with respect to all directions ξα have an eigenvalue of multiplicity n−1 and that for each ξα the distinguished eigendirection is the same [35].

Suppose now that L is an r-dimensional subspace of TpM, r≥2 and let {e1,…,er} be an orthonormal basis of L. Then the scalar curvature τ(L) of the r-plane section L is given by

$$\tau(L) = \sum_{1\leq\alpha<\beta\leq r} K(e_\alpha \wedge e_\beta)$$

and the Casorati curvature C(L) of the subspace L is defined as

$$C(L) = \frac{1}{r}\sum_{\alpha=n+1}^{m}\sum_{i,j=1}^{r}\left(h_{ij}^{\alpha}\right)^2.$$

The normalized δ-Casorati curvature δc(n−1) and δ^c(n−1) are given by

$$\left[\delta_c(n-1)\right]_p = \frac{1}{2}C_p + \frac{n+1}{2n(n-1)}\inf\{C(L)|L \text{ a hyperplane of } T_pM\}$$

and

$$\left[\hat{\delta}_c(n-1)\right]_p = 2C_p - \frac{2n-1}{2n}\sup\{C(L)|L \text{ a hyperplane of } T_pM\}.$$

2.2 Quaternionic Kähler manifolds

We give in this subsection a quick review of basic definitions and properties of manifolds endowed with quaternionic Kähler structures, using mainly [36].

Let M¯ be a differentiable manifold and assume that there is a rank 3-subbundle σ of End(TM¯) such that a local basis {J1,J2,J3} exists on sections of σ satisfying for all α∈{1,2,3}:

$$J_\alpha^2 = -\,\mathrm{Id}, \qquad J_\alpha J_{\alpha+1} = -J_{\alpha+1}J_\alpha = J_{\alpha+2},$$

where Id denotes the identity tensor field of type (1,1) on M and the indices are taken from {1,2,3} modulo 3. Then the bundleσ is called an almost quaternionic structure on *M* and {J1,J2,J3} is called a canonical local basis of σ. Moreover, (M^{-},σ) is said to be an almost quaternionic manifold. It is easy to see that any almost quaternionic manifold is of dimension 4*m*, m≥1.

A Riemannian metric g^{-} on M^{-} is said to be adapted to the almost quaternionic structure σ if it satisfies

$$\bar{g}(J_\alpha X, J_\alpha Y) = \bar{g}(X, Y), \quad \forall \alpha \in \{1,2,3\}$$

for all vector fields *X*, *Y* on M^{-} and any canonical local basis {J1,J2,J3} of σ. Moreover, (M^{-},σ,g^{-}) is said to be an almost quaternionic Hermitian manifold.

If the bundle σ is parallel with respect to the Levi-Civita connection $\nabla^{-}$of g^{-}, then (M^{-},σ,g^{-}) is said to be a quaternionic Kähler manifold. Equivalently, locally defined 1-forms ω1, ω2, ω3 exist such that we have for all α∈{1,2,3}:

$$\overline{\nabla}_X J_\alpha = \omega_{\alpha+2}(X) J_{\alpha+1} - \omega_{\alpha+1}(X) J_{\alpha+2}$$

for any vector field *X* on M^{-}, where the indices are taken from {1,2,3} modulo 3.

We remark that any quaternionic Kähler manifold is an Einstein manifold, provided that dimM>4.

Let (M^{-},σ,g^{-}) be a quaternionic Kähler manifold and let *X* be a non-null vector field on M^{-}. Then the 4-plane spanned by {X,J1X,J2X,J3X}, denoted by Q(X), is called a quaternionic 4-plane. Any 2-plane in Q(X) is called a quaternionic plane. The sectional curvature of a quaternionic plane is called a quaternionic sectional curvature. A quaternionic Kähler manifold is a quaternionic space form if its quaternionic sectional curvatures are equal to a constant, say *c*. It is well known that a quaternionic Kähler manifold (M^{-},σ,g^{-}) is a quaternionic space form, denoted M^{-}(c), if and only if its curvature tensor is given by

$$\overline{R}(X,Y)Z = \frac{c}{4}\Bigg\{\overline{g}(Z,Y)X - \overline{g}(X,Z)Y + \sum_{\alpha=1}^{3}\big[\overline{g}(Z,J_\alpha Y)J_\alpha X - $$

$$- \overline{g}(Z,J_\alpha X)J_\alpha Y + 2\overline{g}(X,J_\alpha Y)J_\alpha Z\big]\Bigg\} \quad (6)$$

for all vector fields *X, Y, Z* on M¯¯¯¯ and any local basis {J1,J2,J3} of *σ*.

A submanifold *M* in a quaternionic Kähler manifold (M¯,σ,g¯) is called a quaternionic submanifold [37] (resp. a totally real submanifold [38]) if each tangent space of *M* is carried into itself (resp. into the normal space) by each section in *σ*. In [39], the author introduced the concept of slant submanifolds as a natural generalization of both quaternionic and totally real submanifolds. A submanifold *M* of a quaternionic Kähler manifold (M¯,σ,g¯) is said to be a slant submanifold if for each non-zero vector *X*tangent to *M* at *p*, the angle θ(X) between Jα(X) and TpM, α∈{1,2,3} is constant, *i.e.* it does not depend on the choice of p∈M and X∈TpM. We can easily see that quaternionic submanifolds are slant submanifolds with θ=0 and totally real submanifolds are slant submanifolds with θ=π2. A slant submanifold of a quaternionic Käler manifold is said to be proper (or *θ*-slant proper) if it is neither quaternionic nor totally real. We note that another natural generalization of both quaternionic and totally real submanifolds in a quaternionic Kähler manifold is given by quaternionic CR-submanifolds. A submanifold *M* of a quaternion Kähler manifold (M¯,σ,g¯) is said to be a quaternionic CR-submanifold if there exist two orthogonal complementary distributions *D* and D⊥ on *M* such that *D* is invariant under quaternionic structure and D⊥ is totally real (see [40]). It is clear that, although quaternionic CR-submanifolds are also the generalization of both quaternionic and totally real submanifolds, there exists no inclusion between the two classes of quaternionic CR-submanifolds and slant submanifolds.

We also note that we have the next characterization of slant submanifolds in quaternionic Kähler manifolds.

Theorem 2.1 [39]*Let M be a submanifold of a quaternionic Kähler manifold* M¯. *Then M is slant if and only if there exists a constant* λ∈[−1,0] *such that*

$$P_\beta P_\alpha X = \lambda X, \quad \forall X \in \Gamma(TM), \alpha, \beta \in \{1,2,3\}, \tag{7}$$

where PαX *denote the tangential component of* JαX. *Furthermore, in such a case, if θ is the slant angle of M, then it satisfies* λ=−cos2θ.

From the above theorem it follows easily that

$$g(P_\alpha X, P_\beta Y) = \cos^2\theta g(X,Y) \tag{8}$$

for X,Y∈Γ(TM) and α,β∈{1,2,3}.

Moreover, every proper slant submanifold of a quaternionic Kähler manifold is of even dimension n=2s≥2 and we can choose a canonical orthonormal local frame, called an adapted slant frame, as follows: $\{e_1, e_2 = \sec\theta P_\alpha e_1, \ldots, e_{2s-1}, e_{2s} = \sec\theta P_\alpha e_{2s-1}\}$, where α is 1, 2 or 3 (see [41]).

3 PROOF OF THEOREM 1.1

*(i)*Since M¯4m(c) is a quaternionic space form, from (6) and Gauss equation (5) we can easily obtain:

$$n^2\|H\|^2 = 2\tau(p) + \|h\|^2 - \frac{n(n-1)c}{4} - \frac{3c}{4}\sum_{\beta=1}^{3}\sum_{i,j=1}^{n} g^2(P_\beta e_i, e_j). \tag{9}$$

Choosing now an adapted slant basis

$$\{e_1, e_2 = \sec\theta P_\alpha e_1, \ldots, e_{2s-1}, e_{2s} = \sec\theta P_\alpha e_{2s-1}\}$$

of TpM, p∈M, where 2s=n and making use of (7) and (8), we derive

$$g^2(P_\beta e_i, e_{i+1}) = g^2(P_\beta e_{i+1}, e_i) = \cos^2\theta \quad \text{for } i = 1, 3, \ldots, 2s-1 \tag{10}$$

and

$$g(P_\beta e_i, e_j) = 0 \quad \text{for } (i,j) \notin \{(2l-1, 2l), (2l, 2l-1) | l \in \{1, 2, \ldots, s\}\}. \tag{11}$$

From (9), (10), and (11) we deduce that

$$2\tau(p) = n^2 \|H\|^2 - n\mathcal{C} + \frac{c}{4}\left[n(n-1) + 9n\cos^2\theta\right]. \tag{12}$$

We define now the following function, denoted by $\mathcal{P}$, which is a quadratic polynomial in the components of the second fundamental form:

$$\mathcal{P} = \frac{1}{2}n(n-1)\mathcal{C} + \frac{1}{2}(n+1)\mathcal{C}(L) - 2\tau(p) + \frac{c}{4}\left[n(n-1) + 9n\cos^2\theta\right],$$

where L is a hyperplane of TpM. We can assume without loss of generality that L is spanned by e1,...,en−1. Then we have

$$\mathcal{P} = \frac{n-1}{2}\sum_{\alpha=n+1}^{4m}\sum_{i,j=1}^{n}(h_{ij}^{\alpha})^2 + \frac{n+1}{2(n-1)}\sum_{\alpha=n+1}^{4m}\sum_{i,j=1}^{n-1}(h_{ij}^{\alpha})^2 - 2\tau(p) + \frac{c}{4}\left[n(n-1) + 9n\cos^2\theta\right]. \tag{13}$$

From (12) and (13), we derive

$$\mathcal{P} = \frac{n+1}{2}\sum_{\alpha=n+1}^{4m}\sum_{i,j=1}^{n}(h_{ij}^{\alpha})^2 + \frac{n+1}{2(n-1)}\sum_{\alpha=n+1}^{4m}\sum_{i,j=1}^{n-1}(h_{ij}^{\alpha})^2 - \sum_{\alpha=n+1}^{4m}\left(\sum_{i=1}^{n}h_{ii}^{\alpha}\right)^2$$

and now we obtain easily that

$$\mathcal{P} = \sum_{\alpha=n+1}^{4m} \sum_{i=1}^{n-1} \left[\frac{n^2 - n + 2}{2(n-1)} (h_{ii}^{\alpha})^2 + (n+1)(h_{in}^{\alpha})^2 \right]$$

$$+ \sum_{\alpha=n+1}^{4m} \left[\frac{n(n+1)}{n-1} \sum_{i<j=1}^{n-1} (h_{ij}^{\alpha})^2 - 2 \sum_{i<j=1}^{n} h_{ii}^{\alpha} h_{jj}^{\alpha} + \frac{n-1}{2} (h_{nn}^{\alpha})^2 \right]. \quad (14)$$

From (14) it follows that the critical points

$$h^c = (h_{11}^{n+1}, h_{12}^{n+1}, \ldots, h_{nn}^{n+1}, \ldots, h_{11}^{4m}, h_{12}^{4m}, \ldots, h_{nn}^{4m})$$

of $\mathcal{P}$ are the solutions of the following system of linear homogeneous equations:

$$\begin{cases} \frac{\partial \mathcal{P}}{\partial h_{ii}^{\alpha}} = \frac{n(n+1)}{n-1} h_{ii}^{\alpha} - 2\sum_{k=1}^{n} h_{kk}^{\alpha} = 0, \\ \frac{\partial \mathcal{P}}{\partial h_{nn}^{\alpha}} = (n-1)h_{nn}^{\alpha} - 2\sum_{k=1}^{n-1} h_{kk}^{\alpha} = 0, \\ \frac{\partial \mathcal{P}}{\partial h_{ij}^{\alpha}} = \frac{2n(n+1)}{n-1} h_{ij}^{\alpha} = 0, \\ \frac{\partial \mathcal{P}}{\partial h_{in}^{\alpha}} = 2(n+1)h_{in}^{\alpha} = 0, \end{cases} \quad (15)$$

with i,j∈{1,...,n−1}, i≠j, and α∈{n+1,...,4m}.

From (15) it follows that every solutions hc has hαij=0 for i≠j and the determinant which corresponds to the first two sets of equations of the above system is zero (there exist solutions for non-totally geodesic submanifolds). Moreover, the Hessian matrix of $\mathcal{P}$ has the eigenvalues

$$\lambda_{11} = 0, \qquad \lambda_{22} = \frac{2n^2 - 5n + 5}{n-1}, \qquad \lambda_{33} = \cdots = \lambda_{nn} = \frac{n(n+1)}{n-1},$$

$$\lambda_{ij} = \frac{2n(n+1)}{n-1}, \qquad \lambda_{in} = 2(n+1), \quad \forall i,j \in \{1,\ldots,n-1\}, i \neq j.$$

Thus, it follows that $\mathcal{P}$ is parabolic and reaches a minimum P(hc) for each solution hc of the system (15). But inserting (15) in (14) we obtain P(hc)=0. So P≥0, and this implies

$$2\tau(p) \le \frac{1}{2}n(n-1)\mathcal{C} + \frac{1}{2}(n+1)\mathcal{C}(L) + \frac{c}{4}\left[n(n-1) + 9n\cos^2\theta\right].$$

Hence we deduce that

$$\rho \le \frac{1}{2}\mathcal{C} + \frac{n+1}{2n(n-1)}\mathcal{C}(L) + \frac{c}{4}\left[1 + \frac{9}{n-1}\cos^2\theta\right]$$

for every tangent hyperplane *L* of *M*. Taking now the infimum over all tangent hyperplane *L* we obtain (1).

Moreover, we can easily see now that the equality sign holds in the inequality (1) if and only if

$$h_{ij}^{\alpha} = 0, \quad \forall i,j \in \{1,\dots,n\}, i \neq j \text{ and } \alpha \in \{n+1,\dots,4m\} \tag{16}$$

and

$$h_{nn}^{\alpha} = 2h_{11}^{\alpha} = 2h_{22}^{\alpha} = \cdots = 2h_{n-1,n-1}^{\alpha}, \quad \forall \alpha \in \{n+1,\dots,4m\}. \tag{17}$$

From (16) and (17) we conclude that the equality sign holds in the inequality (1) if and only if the submanifold *M* is invariantly quasi-umbilical with trivial normal connection in M¯¯¯¯, such that with respect to suitable orthonormal tangent and normal orthonormal frames, the shape operators take the forms (2).

*(ii)*can be proved in a similar way, considering the following quadratic polynomial in the components of the second fundamental form:

$$\mathcal{Q} = 2n(n-1)\mathcal{C} - \frac{1}{2}(2n-1(n-1))\mathcal{C}(L) - 2\tau(p) + \frac{c}{4}\left[n(n-1) + 9n\cos^2\theta\right],$$

where L is a hyperplane of TpM.

ACKNOWLEDGEMENTS

The third author was supported by CNCS-UEFISCDI, project number PN-II-ID-PCE-2011-3-0118.

COMPETING INTERESTS

The authors declare that they have no competing interests.

AUTHORS' CONTRIBUTIONS

All authors contributed equally and significantly in writing this paper. All authors read and approved the final manuscript.

REFERENCES

1. Chen B-Y: **Some pinching and classification theorems for minimal submanifolds.** *Arch. Math.* 1993, **60:** 568-578.

2. Chen B-Y: **δ -Invariants, inequalities of submanifolds and their applications.** In *Topics in Differential Geometry*. Ed. Acad. Române, Bucharest; 2008:29-155.

3. Deng S: **Improved Chen-Ricci inequality for Lagrangian submanifolds in quaternion space forms.** *Int. Electron. J. Geom.*2012,**5**(1):163-170.

4. Hong Y, Houh CS: **Lagrangian submanifolds of quaternion Kaehlerian manifolds satisfying Chen's equality.** *Beitr. Algebra Geom.*1998,**39**(2):413-421.

5. Liu X: **On Ricci curvature of totally real submanifolds in a quaternion projective space.** *Arch. Math.* 2002,**38**(4):297-305.

6. Liu X, Dai W: **Ricci curvature of submanifolds in a quaternion projective space.** *Commun. Korean Math. Soc.* 2002,**17**(4):625-633.

7. Mihai I, Al-Solamy FR, Shahid MH: **On Ricci curvature of a quaternion CR-submanifold in a quaternion space form.** *Rad. Mat.*2003,**12**(1):91-98

8. Shahid MH, Al-Solamy FR: **Ricci tensor of slant submanifolds in a quaternion projective space.** *C. R. Math. Acad. Sci. Paris*2011,**349**(9):571-573.

9. Shukla SS, Rao PK: **Ricci curvature of quaternion slant submanifolds in quaternion space forms.** *Acta Math. Acad. Paedagog. Nyházi* 2012,**28**(1):69-81.

10. Vîlcu GE: **Slant submanifolds of quaternionic space forms.** *Publ. Math. (Debr.)* 2012,**81**(3-4):397-413.

11. Yoon DW: **A basic inequality of submanifolds in quaternionic space forms.** *Balk. J. Geom. Appl.* 2004,**9**(2):92-102.

12. Vîlcu GE: **On Chen invariant and inequalities in quaternionic geometry.** *J. Inequal. Appl.* 2013., **2013:** Article ID 66

13. Alegre P, Chen B-Y, Munteanu M: **Riemannian submersions, δ-invariants, and optimal inequality.** *Ann. Glob. Anal. Geom.*2012,**42**(3):317-331.

14. Al-Solamy FR, Chen B-Y, Deshmukh S: **Two optimal inequalities for anti-holomorphic submanifolds and their applications.** *Taiwan. J. Math.* 2014,**18**(1):199-217.

15. Chen B-Y: **An optimal inequality for CR-warped products in complex space forms involving CR δ -invariant.** *Int. J. Math.*2012.,**23**(3): Article ID 1250045

16. Chen B-Y, Dillen F, Van der Veken J, Vrancken L: **Curvature inequalities for Lagrangian submanifolds: the final solution.** *Differ. Geom. Appl.* 2013,**31**(6):808-819.

17. Chen B-Y, Prieto-Martín A, Wang X:**Lagrangian submanifolds in complex space forms satisfying an improved equality involving** δ(2,2). *Publ. Math. (Debr.)* 2013,**82**(1):193-217.

18. Gülbahar M, Kiliç E, Keleş S: **Chen-like inequalities on lightlike hypersurfaces of a Lorentzian manifold.** *J. Inequal. Appl.* 2013.,**2013:** Article ID 266

19. Gülbahar M, Kiliç E, Keleş S: **Some inequalities on screen homotethic lightlike hypersurfaces of a Lorentzian manifold.** *Taiwan. J. Math.* 2013,**17**(6):2083-2100.

20. Özgür C, Mihai A: **Chen inequalities for submanifolds of real space forms with a semi-symmetric non-metric connection.** *Can. Math. Bull.* 2012,**55**(3):611-622.

21. Al-Solamy FR, Khan MA: **Semi-invariant warped product submanifolds of almost contact manifolds.** *J. Inequal. Appl.* 2012., **2012:**Article ID 127

22. Al-Solamy FR, Khan MA: **Warped product submanifolds of Riemannian product manifolds.** *Abstr. Appl. Anal.* 2012., **2012:** Article ID 724898

23. Al-Solamy FR, Khan MA: **Semi-slant warped product submanifolds of a Kenmotsu manifold.** *Math. Probl. Eng.* 2012., **2012:** Article ID 708191

24. Khan MA, Uddin S, Sachdeva R: **Semi-invariant warped product submanifolds of cosymplectic manifolds.** *J. Inequal. Appl.* 2012.,**2012:** Article ID 19

25. Uddin S, Khan KA: **An inequality for contact CR-warped product submanifolds of nearly cosymplectic manifolds.** *J. Inequal. Appl.*2012., **2012:** Article ID 304

26. Casorati F: **Mesure de la courbure des surfaces suivant l'idée commune. Ses rapports avec les mesures de courbure gaussienne et moyenne.** *Acta Math.* 1890,**14**(1):95-110.

27. Albertazzi L: *Handbook of Experimental Phenomenology: Visual Perception of Shape, Space and Appearance.* Wiley, Chichester; 2013

28. Haesen S, Kowalczyk D, Verstraelen L: **On the extrinsic principal directions of Riemannian submanifolds.** *Note Mat.* 2009,**29**(2):41-53.

29. Verstraelen L: **The geometry of eye and brain.** *Soochow J. Math.* 2004,**30**(3):367-376.

30. Verstraelen L: **Geometry of submanifolds I. The first Casorati curvature indicatrices.** *Kragujev. J. Math.* 2013,**37**(1):5-23.

31. Decu S, Haesen S, Verstraelen L: **Optimal inequalities involving Casorati curvatures.** *Bull. Transylv. Univ. Braşov, Ser. B*2007,**14**(49):85-93. suppl.

32. Decu S, Haesen S, Verstraelen L: **Optimal inequalities characterising quasi-umbilical submanifolds.** *J. Inequal. Pure Appl. Math.*2008.,**9**(3): Article ID 79

33. Ghişoiu V: **Inequalities for the Casorati curvatures of slant submanifolds in complex space forms.** In *Riemannian Geometry and Applications. Proceedings RIGA 2011*. Ed. Univ. Bucureşti, Bucharest; 2011:145-150.

34. Chen B-Y: *Pseudo-Riemannian Geometry, δ-Invariants and Applications*. World Scientific, Hackensack; 2011.

35. Blair D: **Quasi-umbilical, minimal submanifolds of Euclidean space.** *Simon Stevin* 1977, **51:** 3-22.

36. Ishihara S: **Quaternion Kählerian manifolds.** *J. Differ. Geom.* 1974, **9:** 483-500.

37. Gray A:**A note on manifolds whose holonomy group is a subgroup of** Sp(n)·Sp(1). *Mich. Math. J.* 1969, **16:** 125-128.

38. Chen B-Y, Houh CS: **Totally real submanifolds of a quaternion projective space.** *Ann. Mat. Pura Appl.* 1979,**120**(1):185-199.

39. Şahin B: **Slant submanifolds of quaternion Kaehler manifolds.** *Commun. Korean Math. Soc.* 2007,**22**(1):123-135.

40. Barros M, Chen B-Y, Urbano F: **Quaternion CR-submanifolds of quaternion manifolds.** *Kodai Math. J.* 1981, **4:** 399-417.

41. Vîlcu GE: **B.-Y. Chen inequalities for slant submanifolds in quaternionic space forms.** *Turk. J. Math.* 2010,**34**(1):115-128.

7

REMARKS ON INEQUALITIES FOR THE CASORATI CURVATURES OF SLANT SUBMANIFOLDS IN QUATERNIONIC SPACE FORMS

Pan Zhang and Liang Zhang*

School of Mathematics and Computer Science, Anhui Normal University, Huajing South Road, Yijiang District, Wuhu, Anhui 241000, P.R. China

ABSTRACT

In this paper, we obtain an inequality for the normalized Casorati curvature of slant submanifolds in quaternionic space forms by using T Oprea's optimization method.

KEYWORDS

Inequalities, Casorati curvatures, quaternionic space forms

1 INTRODUCTION

The Casorati curvature of an n-dimensional submanifold M of a Riemannian manifold, usually denoted by C, is an extrinsic invariant defined as the normalized square of the length of the second fundamental form of the submanifold. In [1], Decu *et al.*introduced the normalized δ-Casorati curvatures δc(n−1) and δ^c(n−1) by

$$[\delta_c(n-1)]_x = \frac{1}{2}C_x + \frac{n+1}{2n(n-1)}\inf\{C(L) \mid L \text{ a hyperplane of } T_xM\} \tag{1}$$

and

$$[\hat{\delta}_c(n-1)]_x = 2C_x - \frac{2n-1}{2n}\sup\{C(L) \mid L \text{ a hyperplane of } T_xM\},$$

where x∈M, and established some inequalities involving these invariants for submanifolds in real space forms. Later, Slesar *et al.* proved two inequalities relating the above normalized Casorati curvatures for a slant submanifolds in a quaternionic space form in [2]. However, it was pointed out that the coefficient n+12n(n−1) in (1) is inappropriate and must be replaced by n+12n [3, 4]. Following [3, 4], we define the normalized δ-Casorati curvature δC(n−1) by

$$[\delta_C(n-1)]_x = \frac{1}{2}C_x + \frac{n+1}{2n}\inf\{C(L) \mid L \text{ a hyperplane of } T_xM\}. \tag{2}$$

By using T Oprea's optimization method on Riemannian submanifolds, we establish the following inequalities in terms of δC(n−1) for θ-slant proper submanifolds of a quaternionic space form.

Theorem 1 *Let* Mn, n≥3, *be θ-slant proper submanifold of a quaternionic space form* M¯4m(c). *Then the normalized δ-Casorati curvature* δC(n−1) *satisfies*

$$\rho \leq \delta_C(n-1) + \frac{c}{4}\left(1 + \frac{9}{n-1}\cos^2\theta\right), \tag{3}$$

where ρ is the normalized scalar curvature of Mn. *Moreover, the equality case holds if and only if* Mn *is an invariantly quasi-umbilical submanifold with trivial normal connection in* M¯4m(c), *such that with respect to suitable orthonormal tangent frame* {ξ1,...,ξn} *and normal orthonormal frame* {ξn+1,...,ξ4m}, *the shape operators* Ar=Aer, r∈{n+1,...,4m}, *take the following forms*:

$$A_{n+1} = \begin{pmatrix} a & 0 & 0 & \cdots & 0 & 0 \\ 0 & a & 0 & \cdots & 0 & 0 \\ 0 & 0 & a & \cdots & 0 & 0 \\ \vdots & \vdots & \vdots & \ddots & \vdots & \vdots \\ 0 & 0 & 0 & \cdots & a & 0 \\ 0 & 0 & 0 & \cdots & 0 & 2a \end{pmatrix}, \qquad A_{n+2} = \cdots = A_{4m} = 0.$$

2 PRELIMINARIES

Let (Mn,g) be an *n*-dimensional submanifold in an (n+p)-dimensional Riemannian manifold (M¯n+p,g¯). The Levi-Civita connections on M¯¯n+p and Mn will be denoted by ∇¯ and ∇, respectively. For all X,Y∈C∞(TM), N∈C∞(TM⊥), the Gauss and Weingarten formulas can be expressed by

$$\overline{\nabla}_X Y = \nabla_X Y + h(X, Y), \qquad \overline{\nabla}_X N = -A_N X + \nabla_X^{\perp} N,$$

where *h* is the second fundamental form of *M*, ∇¯ is the normal connection and the shape operator AN of *M* is given by

$$g(A_N X, Y) = \overline{g}\big(h(X, Y), N\big).$$

The submanifold *M* is said to be totally geodesic if h=0. Besides, *M* is called invariantly quasi-umbilical if there exist *p* mutually orthogonal unit normal vectors ξn+1,...,ξn+p such that the shape operators with respect to all directions ξr have an eigenvalue of multiplicity n−1 and that for each ξr the distinguished eigendirection is the same [1–4].

In M¯n+p we choose a local orthonormal frame e1,...,en,en+1,...,en+p, such that, restricting ourselves to Mn, e1,...,enare tangent to Mn. We write hrij=g(h(ei,ej),er). Then the *mean curvature vector H* is given by

$$H = \sum_{r=n+1}^{n+p} \left(\frac{1}{n} \sum_{i=1}^{n} h_{ii}^{r} \right) e_r,$$

and the squared norm of *h* over dimension *n* is denoted by C and is called the Casorati curvature of the submanifold *M*. Therefore we have

$$C = \frac{1}{n} \sum_{r=n+1}^{n+p} \sum_{i,j=1}^{n} (h_{ij}^{r})^2.$$

Let K(ei∧ej), 1≤i<j≤n, denote the *sectional curvature* of the plane section spanned by ei and ej. Then the *scalar curvature* of Mn is given by

$$\tau = \sum_{i<j} K(e_i \wedge e_j),$$

and the normalized scalar curvature ρ is defined by

$$\rho = \frac{2\tau}{n(n-1)}.$$

Suppose *L* is an *l*-dimensional subspace of TxM, x∈M, l≥2 and {e1,...,el} an orthonormal basis of *L*. Then the *scalar curvature* τ(L) of the *l*-plane *L* is given by

$$\tau(L) = \sum_{1 \leq \mu < \nu \leq l} K(e_\mu \wedge e_\nu),$$

and the Casorati curvature C(L) of the subspace L is defined as

$$C(L) = \frac{1}{r} \sum_{r=n+1}^{n+p} \sum_{i,j=1}^{n} (h_{ij}^r)^2.$$

For more details of slant submanifolds in quaternionic space forms, we refer to [2, 4].

3 OPTIMIZATION METHOD ON RIEMANNIAN SUBMANIFOLDS

Let (N2,g¯) be a Riemannian manifold, N1 be a Riemannian submanifold of it, g be the metric induced on N1 by g¯ and f:N1→R be a differentiable function.

Following [5–7] we considered the constrained extremum problem

$$\min_{x \in N_1} f(x), \tag{4}$$

then we have the following.

Lemma 1 ([5])

If x0∈N1 *is the solution of the problem* (4), *then*

(i)(gradf)(x0)∈T⊥x0N1;

(ii)the bilinear form

$$\mathcal{A}: T_{x_0}N_1 \times T_{x_0}N_1 \to \mathbb{R};$$

$$\mathcal{A}(X,Y) = \mathrm{Hess}_f(X,Y) + \bar{g}\big(h(X,Y),(\mathrm{grad} f)(x_0)\big)$$

is positive semidefinite, where h is the second fundamental form of N1 *in* N2.

In [6], the above lemma was successfully applied to improve an inequality relating δ(2) obtained in [8]. Later, Chen extended the improved inequality to the general inequalities involving δ-invariants δ(n1,...,nk) [9]. More details of δ-invariants can be found in [10–15]. Besides, the first author gave another proof of the inequalities relating the normalized δ-Casorati curvature δ^c(n−1)for submanifolds in real space forms by using T Oprea's optimization method [16].

4 PROOF OF THEOREM 1

From the Gauss equation we can easily obtain (see (12) in [2])

$$2\tau = \frac{c}{4}\left[n(n-1) + 9n\cos^2\theta\right] + n^2\|H\|^2 - n\mathcal{C}. \tag{5}$$

We define now the following function, denoted by Q, which is a quadratic polynomial in the components of the second fundamental form:

$$\mathcal{Q} = \frac{1}{2}n(n-1)\mathcal{C} + \frac{1}{2}(n+1)(n-1)\mathcal{C}(L) - 2\tau + \frac{c}{4}\left[n(n-1) + 9n\cos^2\theta\right]. \tag{6}$$

Without loss of generality, by assuming that L is spanned by e1,...,en−1, one gets

$$\mathcal{Q} = \frac{n+1}{2}\sum_{\alpha=n+1}^{4m}\left[\sum_{i,j=1}^{n}(h_{ij}^{\alpha})^2\right] + \frac{n+1}{2}\sum_{\alpha=n+1}^{4m}\left[\sum_{i,j=1}^{n-1}(h_{ij}^{\alpha})^2\right] - \sum_{\alpha=n+1}^{4m}\left(\sum_{i=1}^{n}h_{ii}^{\alpha}\right)^2, \tag{7}$$

here we used (5) and (6).

From (7) we have

$$\begin{aligned}\mathcal{Q} &= \sum_{\alpha=n+1}^{4m}\sum_{i=1}^{n-1}\left[n(h_{ii}^{\alpha})^2 + (n+1)(h_{in}^{\alpha})^2\right] \\ &+ \sum_{\alpha=n+1}^{4m}\left[2(n+1)\sum_{1\le i<j\le n-1}(h_{ij}^{\alpha})^2 - 2\sum_{1\le i<j\le n}h_{ii}^{\alpha}h_{jj}^{\alpha} + \frac{n-1}{2}(h_{nn}^{\alpha})^2\right] \\ &\ge \sum_{\alpha=n+1}^{4m}\sum_{i=1}^{n-1}n(h_{ii}^{\alpha})^2 + \sum_{\alpha=n+1}^{4m}\left[-2\sum_{1\le i<j\le n}h_{ii}^{\alpha}h_{jj}^{\alpha} + \frac{n-1}{2}(h_{nn}^{\alpha})^2\right].\end{aligned} \tag{8}$$

For $\alpha=n+1,\dots,4m$, let us consider the quadratic form

$$f_\alpha : \mathbb{R}^n \to \mathbb{R},$$

$$f_\alpha\left(h_{11}^{\alpha},\dots,h_{nn}^{\alpha}\right) = \sum_{i=1}^{n-1}n(h_{ii}^{\alpha})^2 - 2\sum_{1\le i<j\le n}h_{ii}^{\alpha}h_{jj}^{\alpha} + \frac{n-1}{2}(h_{nn}^{\alpha})^2$$

and the constrained extremum problem

$$\min f_\alpha$$

$$\text{subject to } F : h_{11}^{\alpha} + \cdots + h_{nn}^{\alpha} = k^{\alpha},$$

where k^α is a real constant.

The partial derivatives of the function f_α are

$$\frac{\partial f_\alpha}{\partial h_{11}^{\alpha}} = 2nh_{11}^{\alpha} - 2\sum_{i=2}^{n}h_{ii}^{\alpha}, \tag{9}$$

$$\frac{\partial f_\alpha}{\partial h^\alpha_{22}} = 2nh^\alpha_{22} - 2h^\alpha_{11} - 2\sum_{i=3}^{n} h^\alpha_{ii}, \tag{10}$$

$$\frac{\partial f_\alpha}{\partial h^\alpha_{n-1,n-1}} = 2nh^\alpha_{n-1,n-1} - 2\sum_{i=1}^{n-2} h^\alpha_{ii} - 2h^\alpha_{nn}, \tag{11}$$

$$\frac{\partial f_\alpha}{\partial h^\alpha_{nn}} = -2\sum_{i=1}^{n-1} h^\alpha_{ii} + (n-1)h^\alpha_{nn}. \tag{12}$$

For an optimal solution (hα11,hα22,...,hαnn) of the problem in question, the vector gradfα is normal at *F*, that is, it is collinear with the vector (1,1,...,1). From (9), (10), (11), and (12), it follows that a critical point of the considered problem has the form

$$\left(h^\alpha_{11}, h^\alpha_{22}, \ldots, h^\alpha_{n-1,n-1}, h^\alpha_{nn}\right) = \left(t^\alpha, t^\alpha, \ldots, t^\alpha, 2t^\alpha\right). \tag{13}$$

As $\sum_{i=1}^{n} h^\alpha_{ii} = k^\alpha$, by using (13) we have

$$h^\alpha_{11} = h^\alpha_{22} = \cdots = h^\alpha_{n-1,n-1} = \frac{1}{n+1}k^\alpha, \qquad h^\alpha_{nn} = \frac{2}{n+1}k^\alpha. \tag{14}$$

We fix an arbitrary point x∈*F*. The 2-form A:Tx*F*×Tx*F*→R has the expression

$$\mathcal{A}(X,Y) = \mathrm{Hess} f_\alpha(X,Y) + \left\langle h'(X,Y), (\mathrm{grad} f_\alpha)(x)\right\rangle,$$

where h′ is the second fundamental form of *F* in Rn and ⟨,⟩ is the standard inner-product on Rn. In the standard frame of Rn, the Hessian of fα has the matrix

$$\begin{pmatrix} 2n & -2 & -2 & \cdots & -2 & -2 \\ -2 & 2n & -2 & \cdots & -2 & -2 \\ -2 & -2 & 2n & \cdots & -2 & -2 \\ \vdots & \vdots & \vdots & \ddots & \vdots & \vdots \\ -2 & -2 & -2 & \cdots & 2n & -2 \\ -2 & -2 & -2 & \cdots & -2 & n-1 \end{pmatrix}.$$

As F is totally geodesic in Rn, considering a vector X tangent to F at the arbitrary point x on F, that is, verifying the relation ∑ni=1Xi=0, we have

$$\begin{aligned} \mathcal{A}(X,X) &= (X_1, X_2, X_3, \ldots, X_{n-1}, X_n) \begin{pmatrix} 2n & -2 & -2 & \cdots & -2 & -2 \\ -2 & 2n & -2 & \cdots & -2 & -2 \\ -2 & -2 & 2n & \cdots & -2 & -2 \\ \vdots & \vdots & \vdots & \ddots & \vdots & \vdots \\ -2 & -2 & -2 & \cdots & 2n & -2 \\ -2 & -2 & -2 & \cdots & -2 & n-1 \end{pmatrix} \begin{pmatrix} X_1 \\ X_2 \\ X_3 \\ \vdots \\ X_{n-1} \\ X_n \end{pmatrix} \\ &= 2(n+1)\sum_{i=1}^{n-1} X_i^2 + (n+1)X_n^2 - 2(X_1 + X_2 + \cdots + X_n)^2 \\ &= 2(n+1)\sum_{i=1}^{n-1} X_i^2 + (n+1)X_n^2 \\ &\geq 0. \end{aligned}$$

Thus the point (hα11,hα22,...,hαnn) given by (14) is a global minimum point, here we used Lemma 1. Inserting (14) in (8) we have

$$\mathcal{Q} \geq 0. \tag{15}$$

From (2), (6), and (15) we can derive inequality (3). The equality case of (3) holds if and only if we have the equality in all the previous inequalities. Thus

$$h_{ij}^{\alpha} = 0, \quad i \neq j, \forall \alpha;$$

$$h_{nn}^{\alpha} = 2h_{11}^{\alpha} = 2h_{22}^{\alpha} = \cdots = 2h_{n-1,n-1}^{\alpha}, \quad \forall \alpha.$$

ACKNOWLEDGEMENTS

We would like to thank to Professor Weidong Song, who has always been generous with his time and advice.

COMPETING INTERESTS

The authors declare that they have no competing interests.

AUTHORS' CONTRIBUTIONS

All authors contributed equally and significantly in writing this paper. All authors read and approved the final manuscript.

REFERENCES

1. Decu S, Haesen S, Verstraelen L: **Optimal inequalities involving Casorati curvatures.** *Bull. Transilv. Univ. Brasov Ser. B*2007, **14**(49):85–93. suppl.

2. Slesar V, Sahin B, Vîlcu GE: **Inequalities for the Casorati curvatures of slant submanifolds in quaternionic space forms.** *J. Inequal. Appl.* 2014., **2014:** Article ID 123

3. Lee CW, Yoon DW, Lee JW: **Optimal inequalities for the Casorati curvatures of submanifolds of real space forms endowed with semi-symmetric metric connections.** *J. Inequal. Appl.* 2014., **2014:** Article ID 327

4. Lee, JW, Vîlcu, GE: Inequalities for generalized normalized δ-Casorati curvatures of slant submanifolds in quaternionicspace forms (2014). arXiv: 1405.5192v1

5. Oprea T: **Optimization methods on Riemannian submanifolds.** *An. Univ. Bucur., Mat.* 2005, **54:** 127–136.

6. Oprea T: **Chen's inequality in the Lagrangian case.** *Colloq. Math.* 2007, **108:** 163–169. 10.4064/cm108-1-15

7. Oprea T: **Ricci curvature of Lagrangian submanifolds in complex space forms.** *Math. Inequal. Appl.* 2010,**13**(4):851–858.

8. Chen B-Y, Dillen F, Verstraelen L, Vrancken L:**Totally real submanifolds of** CPn **satisfying a basic equality.** *Arch. Math.* 1994, **63:**553–564.

9. Chen B-Y, Dillen F: **Optimal general inequalities for Lagrangian submanifolds in complex space forms.** *J. Math. Anal. Appl.*2011,**379**(1):229–239.

10. Chen B-Y: **Some pinching and classification theorems for minimal submanifolds.** *Arch. Math.* 1993, **60:** 568–578.

11. Al-Solamy FR, Chen B-Y, Deshmukh S: **Two optimal inequalities for anti-holomorphic submanifolds and their applications.** *Taiwan. J. Math.* 2014, **18:** 199–217.

12. Vîlcu GE: **On Chen invariant and inequalities in quaternionic geometry.** *J. Inequal. Appl.* 2013., **2013:** Article ID 66

13. Gülbahar M, Kılıç E, Keles S: **Chen-like inequalities on lightlike hypersurfaces of a Lorentzian manifold.** *J. Inequal. Appl.* 2013.,**2013:** Article ID 266

14. Zhang P, Zhang L, Song W: **Chen's inequalities for submanifolds of a Riemannian manifold of quasi-constant curvature with a semi-symmetric metric connection.** *Taiwan. J. Math.* 2014.

15. Zhang P: **Remarks on Chen's inequalities for submanifolds of a Riemannian manifold of nearly quasi-constant curvature.** *Vietnam J. Math.* 2014.

16. Zhang, P: Inequalities for Casorati curvatures of submanifolds in real space forms (2014). arXiv: 1408.4996[math.DG]

8

BOUNDARY VALUE PROBLEMS FOR THE QUATERNIONIC HERMITIAN SYSTEM IN R4*N*

Ricardo Abreu-Blaya1 , Juan Bory-Reyes2 , Fred Brackx3 , Hennie De Schepper3* and Frank Sommen3

1Facultad de Informática y Matemática, Universidad de Holguín, Holguín 80100, Cuba.

2Departamento de Matemática, Universidad de Oriente, Santiago de Cuba 90500, Cuba.

3Department of Mathematical Analysis, Faculty of Engineering, Ghent University, Galglaan 2, 9000 Gent, Belgium.

ABSTRACT

In this paper boundary value problems for quaternionic Hermitian monogenic functions are presented using a circulant matrix approach.

KEYWORDS

Quaternionic, Clifford analysis, Cauchy integral formula

1 INTRODUCTION

Euclidean Clifford analysis is a higher dimensional function theory offering a refinement of classical harmonic analysis. The theory is centred around the concept of monogenic functions, *i.e.* null solutions of a first-order vector-valued rotation invariant differential operator, called Dirac operator, which factorises the Laplacian; monogenic functions may thus also be seen as a generalisation of holomorphic functions in the complex plane. Its roots go back to the 1930s. For more details on this function theory we refer to the standard references [5, 12, 14–16].

More recently Hermitian Clifford analysis emerged as a refinement of the Euclidean setting for the case of $\mathbb{R}^{2n}$. Here, Hermitian monogenic functions are considered, *i.e.* functions taking values either in a complex Clifford algebra or in complex spinor space, and being simultaneous null solutions of two complex Hermitian Dirac operators, which are invariant under the action of the unitary group. For the systematic development of this function theory we refer to [6–8].

In the papers [10, 11, 13, 17], the Hermitian Clifford analysis setting was further refined by considering functions on R4n$\mathbb{R}^{4n}$ with values in a quaternionic Clifford algebra, being simultaneous null solutions of four mutually related quaternionic Dirac operators, which are invariant under the action of the symplectic group. In [3], Borel-Pompeiu and Cauchy integral formulas are established in this setting, by following a (4×4) circulant matrix approach, similar in spirit to the circulant (2×2) matrix approach introduced in [9] within the complex Hermitian Clifford case. Subsequently, in [4] a quaternionic Hermitian Cauchy integral is introduced, as well as its boundary limit values, leading to the definition of a matrix quaternionic Hermitian Hilbert transform. These operators provide a useful tool for studying boundary value problems for the quaternionic Hermitian system. This is precisely the main objective of the present paper. The main problems that we address are the problem of finding a quaternionic Hermitian monogenic function with a given jump over a given surface of R4n$\mathbb{R}^{4n}$ as well as problems of Dirichlet type for the quaternionic Hermitian system. Finally, we also prove an equivalence between both-sided quaternionic Hermitian monogenicity and a certain integral conservation law.

2 PRELIMINARIES

Let (e1,...,em) be an orthonormal basis of Euclidean space Rm and consider the real Clifford algebra R0,m constructed over Rm. The non-commutative multiplication in Rm$\mathbb{R}_m$ is governed by the rules:

$$e_\ell^2 = -1, \quad \ell = 1,\ldots,m,$$

$$e_\ell e_k + e_k e_\ell = 0, \quad \ell \neq k.$$

In $\mathbb{R}_m$ one can consider the following automorphisms:

*(i)*the conjugation e¯ℓ=−eℓ and for any a,b∈Rm, ab¯=b¯a¯

*(ii)*the main involution e˜ℓ=−eℓ and for any a,b∈Rm, ab˜=a˜b˜.

In particular, we consider the skew-field of quaternions H$\mathbb{H}$ whose elements will be denoted by q=x0+ix1+jx2+kx3 with i2=j2=k2=−1 and ij=−ji=k. Clearly, H$\mathbb{H}$ may be identified with the Clifford algebra R0,2 making the identifications i↔e1, j↔e2 and k↔e1e2. The automorphisms (i) and (ii) then respectively lead to the H$\mathbb{H}$-conjugation

$$\overline{q} = x_0 - ix_1 - jx_2 - kx_3$$

and to the main H$\mathbb{H}$-involution

$$q^\gamma \equiv \tilde{q} = x_0 - ix_1 - jx_2 + kx_3.$$

However, it is quite natural to introduce two more HH-involutions defined by

$$q^{\alpha} = x_0 + ix_1 - jx_2 - kx_3, \qquad q^{\beta} = x_0 - ix_1 + jx_2 - kx_3.$$

Definition 1 (see[17]) The quaternionic Witt basis of $\mathbb{H}_m = \mathbb{H} \otimes_{\mathbb{R}} \mathbb{R}_m$, $m = 4n$, is given by $\{f_\ell, f_\ell^{\alpha}, f_\ell^{\beta}, f_\ell^{\gamma}\}, \ell = 1, \ldots,$, $\ell = 1, \ldots, n$, where

$$f_\ell = e_{1+4(\ell-1)} - ie_{2+4(\ell-1)} - je_{3+4(\ell-1)} - ke_{4+4(\ell-1)},$$

$$f_\ell^{\alpha} = e_{1+4(\ell-1)} - ie_{2+4(\ell-1)} + je_{3+4(\ell-1)} + ke_{4+4(\ell-1)},$$

$$f_\ell^{\beta} = e_{1+4(\ell-1)} + ie_{2+4(\ell-1)} - je_{3+4(\ell-1)} + ke_{4+4(\ell-1)},$$

$$f_\ell^{\gamma} = e_{1+4(\ell-1)} + ie_{2+4(\ell-1)} + je_{3+4(\ell-1)} - ke_{4+4(\ell-1)}.$$

We will consider the Clifford vectors

$$\underline{X} = \underline{X}_0 = \sum_{\ell=1}^{n} (e_{4\ell-3}x_{4\ell-3} + e_{4\ell-2}x_{4\ell-2} + e_{4\ell-1}x_{4\ell-1} + e_{4\ell}x_{4\ell}),$$

$$\underline{X}_1 = \sum_{\ell=1}^{n} (e_{4\ell-3}x_{4\ell-2} - e_{4\ell-2}x_{4\ell-3} - e_{4\ell-1}x_{4\ell} + e_{4\ell}x_{4\ell-1}),$$

$$\underline{X}_2 = \sum_{\ell=1}^{n} (e_{4\ell-3}x_{4\ell-1} + e_{4\ell-2}x_{4\ell} - e_{4\ell-1}x_{4\ell-3} - e_{4\ell}x_{4\ell-2}),$$

$$\underline{X}_3 = \sum_{\ell=1}^{n} (e_{4\ell-3}x_{4\ell} - e_{4\ell-2}x_{4\ell-1} + e_{4\ell-1}x_{4\ell-2} - e_{4\ell}x_{4\ell-3})$$

for which $\underline{X}_r^2 = -|\underline{X}_0|^2$, while $\underline{X}_r\underline{X}_s + \underline{X}_s\underline{X}_r = 0, r \neq s, r, s = 0, \ldots, 3$. The corresponding Dirac operators are denoted by $\partial_{\underline{X}} = \partial_{\underline{X}_0}$, $\partial_{\underline{X}_1}$, $\partial_{\underline{X}_2}$ and $\partial_{\underline{X}_3}$. and $\partial_{\underline{X}_3}$. Here we have $\partial_{\underline{X}_r}^2 = -\Delta_{4n}$, with Δ_{4n} with Δ_{4n} the Laplacian in

R4n $\mathbb{R}^{4n}$, and $\partial_{\underline{X}_r}\partial_{\underline{X}_s} + \partial_{\underline{X}_s}\partial_{\underline{X}_r} = 0,\ r \neq s,\ r,s = 0,\ldots,3.$ Next, the quaternionic Hermitian variables are introduced:

$$\underline{Z} = \underline{Z}_0 = \underline{X}_0 + i\underline{X}_1 + j\underline{X}_2 + k\underline{X}_3,$$

$$\underline{Z}_1 = \underline{X}_0 + i\underline{X}_1 - j\underline{X}_2 - k\underline{X}_3,$$

$$\underline{Z}_2 = \underline{X}_0 - i\underline{X}_1 + j\underline{X}_2 - k\underline{X}_3,$$

$$\underline{Z}_3 = \underline{X}_0 - i\underline{X}_1 - j\underline{X}_2 + k\underline{X}_3$$

for which $\underline{Z}_0\underline{Z}_0^\dagger + \underline{Z}_1\underline{Z}_1^\dagger + \underline{Z}_2\underline{Z}_2^\dagger + \underline{Z}_3\underline{Z}_3^\dagger = 16|\underline{X}|^2,$ the symbol † denoting Hermitian quaternionic conjugation is defined as the composition of H$\mathbb{H}$-conjugation and Clifford conjugation in $\mathbb{R}_{0,m}$, *i.e.* $\lambda^\dagger = \sum_A \overline{e_A}\overline{\lambda_A}$. The Hermitian Dirac operators are

$$\partial_{\underline{Z}_0} = \frac{1}{16}(\partial_{\underline{X}_0} + i\partial_{\underline{X}_1} + j\partial_{\underline{X}_2} + k\partial_{\underline{X}_3}),$$

$$\partial_{\underline{Z}_1} = \frac{1}{16}(\partial_{\underline{X}_0} + i\partial_{\underline{X}_1} - j\partial_{\underline{X}_2} - k\partial_{\underline{X}_3}),$$

$$\partial_{\underline{Z}_2} = \frac{1}{16}(\partial_{\underline{X}_0} - i\partial_{\underline{X}_1} + j\partial_{\underline{X}_2} - k\partial_{\underline{X}_3}),$$

$$\partial_{\underline{Z}_3} = \frac{1}{16}(\partial_{\underline{X}_0} - i\partial_{\underline{X}_1} - j\partial_{\underline{X}_2} + k\partial_{\underline{X}_3})$$

for which $\Delta_{4n} = 16(\partial_{\underline{Z}_0}\partial_{\underline{Z}_0}^\dagger + \partial_{\underline{Z}_1}\partial_{\underline{Z}_1}^\dagger + \partial_{\underline{Z}_2}\partial_{\underline{Z}_2}^\dagger + \partial_{\underline{Z}_3}\partial_{\underline{Z}_3}^\dagger).$ **Definition 2** (see [17])

Let Ω be an open set in R4n. A continuously differentiable function f:Ω↦H4n is said to be (left) *q*-Hermitian monogenic in Ω (or *q*-monogenic for short) iff it satisfies in Ω the system $\partial_{\underline{Z}_0}f = \partial_{\underline{Z}_1}f = \partial_{\underline{Z}_2}f = \partial_{\underline{Z}_3}f = 0,$ or, equivalently, the system $\partial_{\underline{X}_0}f = \partial_{\underline{X}_1}f = \partial_{\underline{X}_2}f = \partial_{\underline{X}_3}f = 0.$

Similarly right *q*-monogenicity is defined. Left and right *q*-monogenic functions are called two-sided *q*-monogenic. A *q*-monogenic function in Ω is monogenic, and thus harmonic in Ω. Note that Definition 2 was proven in [10] to be equivalent to the system introduced in [13] by group invariance considerations.

The fundamental solutions of the Dirac operators $\partial_{\underline{X}_r}$, r=0,...,3, *i.e.* the Euclidean Cauchy kernels, are respectively given by

$$E_r(\underline{X}) = -\frac{1}{a_{4n}} \frac{\underline{X}_r}{|\underline{X}|^{4n}}, \quad r = 0, \ldots, 3$$

with a4na_{4n} the area of the unit sphere S4n−1S^{4n-1} in R4n$\mathbb{R}^{4n}$. Explicitly, this means that $\partial_{\underline{X}_r} E_r(\underline{X}) = \delta(\underline{X})$, $r = 0, \ldots, 3$. Next we introduce the Hermitian Cauchy kernels:

$$\mathcal{E}_r(\underline{Z}) = \frac{1}{a_{4n}} \frac{\underline{Z}_r^{\dagger}}{|\underline{Z}|^{4n}}, \quad r = 0, \ldots, 3.$$

Note that Er$\mathcal{E}_r$ is not the fundamental solution of ∂Z−r. However, the following theorem holds, see [3].

Theorem 1 *Introducing the circulant* (4×4$4 \times 4$) *matrices*

$$\mathcal{D} = \begin{pmatrix} \partial_{\underline{Z}_0} & \partial_{\underline{Z}_3} & \partial_{\underline{Z}_2} & \partial_{\underline{Z}_1} \\ \partial_{\underline{Z}_1} & \partial_{\underline{Z}_0} & \partial_{\underline{Z}_3} & \partial_{\underline{Z}_2} \\ \partial_{\underline{Z}_2} & \partial_{\underline{Z}_1} & \partial_{\underline{Z}_0} & \partial_{\underline{Z}_3} \\ \partial_{\underline{Z}_3} & \partial_{\underline{Z}_2} & \partial_{\underline{Z}_1} & \partial_{\underline{Z}_0} \end{pmatrix}, \qquad \mathcal{E} = \begin{pmatrix} \mathcal{E}_0 & \mathcal{E}_3 & \mathcal{E}_2 & \mathcal{E}_1 \\ \mathcal{E}_1 & \mathcal{E}_0 & \mathcal{E}_3 & \mathcal{E}_2 \\ \mathcal{E}_2 & \mathcal{E}_1 & \mathcal{E}_0 & \mathcal{E}_3 \\ \mathcal{E}_3 & \mathcal{E}_2 & \mathcal{E}_1 & \mathcal{E}_0 \end{pmatrix},$$

$$\delta = \begin{pmatrix} \delta & 0 & 0 & 0 \\ 0 & \delta & 0 & 0 \\ 0 & 0 & \delta & 0 \\ 0 & 0 & 0 & \delta \end{pmatrix}$$

one obtains that DTE=EDT=δ.

Thus, E is a fundamental solution of D, in a matricial interpretation.

We associate, with functions g0, g1, g2 and g3 defined in Ω⊂R4n and taking values in H4n, the (4×44 × 4) circulant matrix function

$$\boldsymbol{G} = \begin{pmatrix} g_0 & g_3 & g_2 & g_1 \\ g_1 & g_0 & g_3 & g_2 \\ g_2 & g_1 & g_0 & g_3 \\ g_3 & g_2 & g_1 & g_0 \end{pmatrix} \equiv \text{circ} \begin{pmatrix} g_0 \\ g_1 \\ g_2 \\ g_3 \end{pmatrix}. \tag{1}$$

We say that ***G*** belongs to some class of functions if all its entries belong to that class. In particular, the spaces of *k*-times continuously differentiable, of *α*-Hölder continuous (0<α≤1) and of *p*-integrable (4×4) circulant matrix functions on some suitable subset **E** of R4n are respectively denoted by Ck(E)$\mathbf{C}^k(\mathbf{E})$, C0,α(E) and Lp(E). The corresponding spaces of H4n-valued functions are denoted by Ck(E), C0,α(E) and Lp(E). Moreover, introducing the non-negative function ‖G(X−−)‖=maxr=0,1,2,3{|gr(X−−)|}, the classes C0,α(E) and Lp(E) may also be defined by means of the respective traditional conditions

$$\|\mathbf{G}\|_\alpha = \max_{\underline{X}\in\mathbf{E}} \left\| \mathbf{G}(\underline{X}) \right\| + \sup_{\underline{X},\underline{Y}\in\mathbf{E},\underline{X}\neq\underline{Y}} \frac{\|\mathbf{G}(\underline{X}) - \mathbf{G}(\underline{Y})\|}{|\underline{X} - \underline{Y}|^\alpha} < +\infty$$

and

$$\|\mathbf{G}\|_p = \left(\int_{\mathbf{E}} \left\| \mathbf{G}(\underline{X}) \right\|^p \right)^{\frac{1}{p}} < +\infty.$$

Definition 3 The (4×44 × 4) circulant matrix function ***G*** is called (left) **Q**-Hermitian monogenic in Ω (or *Q*-monogenic for short) iff DTG=O in Ω, where ***O*** denotes the matrix with zero entries.

Similarly right **Q**-monogenicity is defined by the system GDT=O. Left and right **Q**-monogenic matrix functions are called two-sided **Q**-monogenic. An important special case concerns the diagonal matrix function G0$\mathbf{G}_0$, with g0=g$g_0 = g$ and g1=g2=g3=0. Indeed, G0 is left (respectively right) -monogenic iff the function g is left (respectively right) q-monogenic.

Now, let Ω+=Ω be a bounded simply connected domain in R4n with boundary Γ=∂Ω, and denoteby Ω^- the complementary open domain R4n\(Ω∪Γ).We assume Γ to be a Liapunov surface. The unit normal vector on Γ at $\underline{X} \in \Gamma$ is given by

$$\underline{n}_0(\underline{X}) = \sum_{\ell=1}^{n} \left(e_{4\ell-3}n_{4\ell-3}(\underline{X}) + e_{4\ell-2}n_{4\ell-2}(\underline{X}) + e_{4\ell-1}n_{4\ell-1}(\underline{X}) + e_{4\ell}n_{4\ell}(\underline{X})\right)$$

and similarly as above, we also introduce the vectors n−1$\underline{n}_1$, n−2$\underline{n}_2$ and n−3$\underline{n}_3$, giving rise in the usual way (up to a constant factor) to their Hermitian counterparts

$$\mathcal{N}_0 = \frac{1}{16}(\underline{n}_0 + i\underline{n}_1 + j\underline{n}_2 + k\underline{n}_3)$$

and N1, N2, N3, as well as to the circulant matrix N$\mathcal{N}$. Then, in [3], the following Cauchy integral formulae were proven for **Q**-monogenic matrix functions and for q-monogenic functions, respectively.

Theorem 2 (**Q**-Hermitian Cauchy integral formula)

If the matrix function ***G****,* (1), *is* **Q***-monogenic in Ω then*

$$\int_{\partial\Gamma} \mathcal{E}(\underline{Z} - \underline{V})\mathcal{N}^T(\underline{Z})\mathbf{G}(\underline{X})\, dS(\underline{X}) = \begin{cases} \mathbf{G}(\underline{Y}), & \underline{Y} \in \Gamma^+, \\ \mathbf{O}, & \underline{Y} \in \Gamma^-. \end{cases}$$

Theorem 3 (q-Hermitian Cauchy integral formula)

If the function g is q-monogenic in Ω then

$$\int_{\partial\Gamma} \mathcal{E}(\underline{Z} - \underline{V})\mathcal{N}^T(\underline{Z})\mathbf{G}_0(\underline{X})\, dS(\underline{X}) = \begin{cases} \mathbf{G}_0(\underline{Y}), & \underline{Y} \in \Gamma^+, \\ \mathbf{O}, & \underline{Y} \in \Gamma^-, \end{cases}$$

where G0$\mathbf{G}_0$ *is the corresponding diagonal matrix.*

Next, in [4] a **Q**-Hermitian Cauchy transform was introduced, given by

$$\mathcal{C}[\mathbf{G}](\underline{Y}) = \int_{\Gamma} \mathcal{E}(\underline{Z} - \underline{V})\mathcal{N}^{T}(\underline{Z})\mathbf{G}(\underline{X})\, dS(\underline{X}), \quad \underline{Y} \notin \Gamma \tag{2}$$

for a matrix function G∈C(Γ), where $\underline{Z}$ and V− denote the Hermitian versions of the Clifford vectors $\underline{X}$ and $\underline{Y}$, respectively. C[G] is a left **Q**-monogenic matrix function in R4n\Γ, vanishing at infinity; in terms of the Euclidean Cauchy type integrals

$$C_{r,s}g(\underline{Y}) := \int_{\Gamma} E_r(\underline{X} - \underline{Y})\underline{n}_s(\underline{X})g(\underline{X})\, dS(\underline{X}), \quad \underline{Y} \notin \Gamma$$

it reads as

$$\mathcal{C}[\mathbf{G}] = \frac{1}{4}\,\mathrm{circ}\begin{pmatrix} C_{0,0} + C_{1,1} + C_{2,2} + C_{3,3} \\ C_{0,0} - C_{2,2} + j(C_{1,3} + C_{3,1}) \\ C_{0,0} - C_{1,1} + C_{2,2} - C_{3,3} \\ C_{0,0} - C_{2,2} - j(C_{1,3} + C_{3,1}) \end{pmatrix}[\mathbf{G}].$$

In particular, for the special case of the matrix G0$\mathbf{G}_0$, the action of C$\mathcal{C}$ is reduced to

$$\mathcal{C}[\mathbf{G}_0] = \frac{1}{4}\,\mathrm{circ}\begin{pmatrix} C_{0,0}g + C_{1,1}g + C_{2,2}g + C_{3,3}g \\ C_{0,0}g - C_{2,2}g + j(C_{1,3}g + C_{3,1}g) \\ C_{0,0}g - C_{1,1}g + C_{2,2}g - C_{3,3}g \\ C_{0,0}g - C_{2,2}g - j(C_{1,3}g + C_{3,1}g) \end{pmatrix}.$$

In general C[G0] will not be a diagonal matrix, whence its entries will not be left q-monogenic functions. However C[G0]$\mathcal{C}[\mathbf{G}_0]$ does become diagonal if and only if

$$C_{0,0}g = C_{2,2}g, \qquad C_{1,3}g = -C_{3,1}g, \qquad 2C_{0,0}g = C_{1,1}g + C_{3,3}g \tag{3}$$

in which case we obtain

$$\mathcal{C}[\mathbf{G}_0] = \operatorname{circ}\begin{pmatrix} C_{0,0}g \\ 0 \\ 0 \\ 0 \end{pmatrix} = \operatorname{circ}\begin{pmatrix} C_{2,2}g \\ 0 \\ 0 \\ 0 \end{pmatrix} = \frac{1}{2}\operatorname{circ}\begin{pmatrix} C_{1,1}g + C_{3,3}g \\ 0 \\ 0 \\ 0 \end{pmatrix}.$$

The following Plemelj-Sokhotski formula, proven in [4], then asserts the existence of the continuous boundary limits of the **Q**-Hermitian Cauchy transform.

Theorem 4 *Let* G∈C0,α(Γ) (0<α≤1), *then the continuous limit values of its* **Q**-*Hermitian Cauchy transform* C[G] *exist and are given by*

$$\mathcal{C}^{\pm}[\mathbf{G}](\underline{U}) = \frac{1}{2}\left(\mathcal{H}[\mathbf{G}](\underline{U}) \pm \mathbf{G}(\underline{U})\right), \quad \underline{U} \in \Gamma.$$

Here we have introduced the matrix **Q**-Hermitian Hilbert operator

$$\mathcal{H}[\mathbf{G}] = \frac{1}{4}\operatorname{circ}\begin{pmatrix} H_{0,0} + H_{1,1} + H_{2,2} + H_{3,3} \\ H_{0,0} - H_{2,2} + j(H_{1,3} + H_{3,1}) \\ H_{0,0} - H_{1,1} + H_{2,2} - H_{3,3} \\ H_{0,0} - H_{2,2} - j(H_{1,3} + H_{3,1}) \end{pmatrix}[\mathbf{G}],$$

where the singular integrals

$$H_{r,s}g(\underline{U}) = 2\int_{\Gamma} E_r(\underline{X} - \underline{U})\underline{n}_s(\underline{X})g(\underline{X})\,dS(\underline{X}), \quad \underline{U} \in \Gamma$$

are Cauchy principal values. H$\mathcal{H}$ shows the following traditional properties, see [4].

Theorem 5 *One has*

(i)$\mathcal{H}$ is a bounded linear operator on (C0,α(Γ),‖·‖α)$(C^{0,\alpha}(\Gamma), \| \bullet \|_\alpha)$ (0<α<1$0 < \alpha < 1$)

(ii)$\mathcal{H}$ is an involution on $C^{0,\alpha}(\Gamma)$ $(0 < \alpha < 1)$.

Similar results may be obtained for right-hand versions of the **Q**-Hermitian Cauchy and Hilbert transforms by means of the alternative definitions

$$[\boldsymbol{G}]\mathcal{C}(\underline{Y}) = \int_\Gamma \mathbf{G}(\underline{X})\mathcal{N}^T(\underline{Z})\mathcal{E}(\underline{Z}-\underline{V})\,dS(\underline{X}), \quad \underline{Y} \notin \Gamma$$

and

$$[\boldsymbol{G}]\mathcal{H} = [\boldsymbol{G}]\frac{1}{4}\operatorname{circ}\begin{pmatrix} H_{0,0}+H_{1,1}+H_{2,2}+H_{3,3} \\ H_{0,0}-H_{2,2}+j(H_{1,3}+H_{3,1}) \\ H_{0,0}-H_{1,1}+H_{2,2}-H_{3,3} \\ H_{0,0}-H_{2,2}-j(H_{1,3}+H_{3,1}) \end{pmatrix},$$

Where

$$gH_{r,s}(\underline{U}) = 2\int_\Gamma g(\underline{X})\underline{n}_s(\underline{X})E_r(\underline{X}-\underline{U})\,dS(\underline{X}), \quad \underline{U} \in \Gamma.$$

3 BOUNDARY VALUE PROBLEMS FOR Q-MONOGENIC FUNCTIONS

In this section we study the so-called jump problem (reconstruction problem) for **Q**-monogenic functions; that is, we will investigate the problem of reconstructing a **Q**-monogenic matrix function $\boldsymbol{\Psi}$ in $\mathbb{R}^{4n} \setminus \Gamma$ vanishing at infinity and having a prescribed jump $\boldsymbol{G}$ across Γ, *i.e.*

$$\mathbf{\Psi}^{+}(\underline{U}) - \mathbf{\Psi}^{-}(\underline{U}) = \mathbf{G}(\underline{U}), \quad \underline{U} \in \Gamma. \tag{4}$$

First, it should be noted that if this problem has a solution, then it necessarily is unique. This assertion can be easily proven using the Painlevé and Liouville theorems in the Clifford analysis setting, see [1]. Next, under the condition that $\mathbf{G} \in C^{0,\alpha}(\Gamma)$, Theorem 4 ensures the solvability of the jump problem (4), its unique solution being given by

$$\mathbf{\Psi}(\underline{Y}) = \mathcal{C}[\mathbf{G}](\underline{Y}), \quad \underline{Y} \in \mathbb{R}^{4n} \setminus \Gamma.$$

Now consider the important special case of the matrix function G0$\mathbf{G}_0$. The reconstruction problem (4) then is strongly related to the jump problem for the involved *q*-monogenic function, as addressed in the following theorem.

Theorem 6 *For a function* $g \in C^{0,\alpha}(\Gamma)$, *the following statements are equivalent*:

(i)the jump problem

$$\psi^{+}(\underline{U}) - \psi^{-}(\underline{U}) = g(\underline{U}), \quad \underline{U} \in \Gamma \tag{5}$$

is solvable in terms of q-monogenic functions;

(ii)g satisfies the relations (3);

(iii)g satisfies the relations C0,0g=C1,1g=C2,2g=C3,3g.

Proof (i) → (ii)

Associate to the function g the diagonal matrix function $\mathbf{G}_0$. Then G0∈C0,α(Γ), and the jump problem (4) for $\mathbf{G}_0$ has the unique solution

$$\mathbf{\Psi}(\underline{Y}) = \mathcal{C}[\mathbf{G}_0](\underline{Y}), \quad \underline{Y} \in \mathbb{R}^{4n} \setminus \Gamma.$$

Let ψ be a solution of (5), then the circulant matrix

$$\mathbf{\Psi}(\underline{Y}) = \text{circ}\begin{pmatrix}\psi\\0\\0\\0\end{pmatrix}$$

is another solution of the jump problem (4) for $\mathcal{G}_0$, whence the uniqueness yields

$$\text{circ}\begin{pmatrix}\psi\\0\\0\\0\end{pmatrix} = \frac{1}{4}\,\text{circ}\begin{pmatrix}C_{0,0}g + C_{1,1}g + C_{2,2}g + C_{3,3}g\\ C_{0,0}g - C_{2,2}g + j(C_{1,3}g + C_{3,1}g)\\ C_{0,0}g - C_{1,1}g + C_{2,2}g - C_{3,3}g\\ C_{0,0}g - C_{2,2}g - j(C_{1,3}g + C_{3,1}g)\end{pmatrix}$$

implying (ii).

(ii)→ (iii)

From the third relation in (3), we have $2\partial_{\underline{Y}_1} C_{0,0}g = \partial_{\underline{Y}_1} C_{3,3}g + \partial_{\underline{Y}_1} C_{1,1}g = \partial_{\underline{Y}_1} C_{3,3}$, and hence

$$\begin{aligned}2\partial_{\underline{Y}_1} C_{0,0}g &= \int_\Gamma \left(\partial_{\underline{Y}_1} E_3(\underline{X} - \underline{Y})\right)\underline{n}_3(\underline{X}) g(\underline{X})\, dS(\underline{X})\\ &= -\int_\Gamma \left(\partial_{\underline{Y}_3} E_1(\underline{X} - \underline{Y})\right)\underline{n}_3(\underline{X}) g(\underline{X})\, dS(\underline{X})\\ &= -\partial_{\underline{Y}_3} C_{1,3}g = \partial_{\underline{Y}_3} C_{3,1}g = 0, \quad \underline{Y} \notin \Gamma\end{aligned}$$

the latter following from the second relation in (3) and the ∂Y−3-monogenicity of C3,1g. This fact means that C0,0g−C3,3g is a ∂Y−1$^{\partial_{\underline{Y}_1}}$-monogenic function in R4n\Γ. Moreover, it has a null jump through Γ, whence it vanishes in the whole of R4n. We conclude that C0,0g=C3,3g. Similarly, we arrive at C0,0g=C1,1g.

(iii)→ (i)

It suffices to observe that, under the conditions stated, $C_{0,0}g$ is *q*-monogenic, whence it solves the jump problem (5). □

For right *q*-monogenic functions the following analogue is obtained.

Theorem 7 *For a function* $g \in C^{0,\alpha}(\Gamma)$, *the following statements are equivalent*:

(i)the jump problem

$$\psi^{+}(\underline{U}) - \psi^{-}(\underline{U}) = g(\underline{U}), \quad \underline{U} \in \Gamma \tag{6}$$

is solvable in terms of right q-monogenic functions;

(ii)g satisfies the relations

$$gC_{0,0} = gC_{2,2}, \qquad gC_{1,3} = -gC_{3,1}, \qquad 2gC_{0,0} = gC_{1,1} + gC_{3,3};$$

(iii)g satisfies the relations $gC_{0,0} = gC_{1,1} = gC_{2,2} = gC_{3,3}$.

The next result deals with the Dirichlet boundary value problem for **Q**-monogenic functions.

Theorem 8 *Let* $\mathbf{G} \in C^{0,\alpha}(\Gamma)$, *then the following statements are equivalent*:

(i)The Dirichlet problem

$$\begin{aligned} &\mathcal{D}^{T}\mathbf{F} = \mathbf{O} \qquad (\textit{resp. } \mathbf{F}\mathcal{D}^{T} = \mathbf{O}), \quad \textit{in } \Omega, \\ &\mathbf{F} = \mathbf{G}, \qquad \textit{on } \Gamma \end{aligned} \tag{7}$$

has a solution.

*(ii)*H[G]=G (*resp.* [G]H=G).

Proof We give the proof for the left-sided version of the theorem, the right-sided one being completely similar.

(i)→ (ii)

Let **F** be a solution of the Dirichlet problem (7). Then, by the **Q**-Hermitian Cauchy formula, we have

$$\mathcal{C}[\mathbf{F}](\underline{Y}) = \mathbf{F}(\underline{Y}), \quad \underline{Y} \in \Omega.$$

Taking limits as Y-→U-∈Γ$\underline{Y} \to \underline{U} \in \Gamma$, (ii) follows in view of Theorem 4.

(ii)→ (i)

It suffices to observe that, under the condition (ii), $\mathbf{F} = \mathcal{C}[\mathbf{G}]$ solves (7).
□

Theorem 9 *Let* $g \in C^{0,\alpha}(\Gamma)$, *then the following statements are equivalent*:

(i)The Dirichlet problem

$$\partial_{\underline{Z}_0} f = \partial_{\underline{Z}_1} f = \partial_{\underline{Z}_2} f = \partial_{\underline{Z}_3} f = 0, \quad in\ \Omega,$$

$$f = g, \quad on\ \Gamma \tag{8}$$

has a solution.

(ii)g satisfies the relations

$$H_{0,0}g = H_{2,2}g = g, \qquad H_{1,3}g = -H_{3,1}g, \qquad H_{1,1}g + H_{3,3}g = 2g.$$

(iii)g satisfies the relations

$$H_{0,0}g = H_{1,1}g = H_{2,2}g = H_{3,3}g = g.$$

Proof (i) → (ii)

From (i) we see that the matrix function

$$\mathbf{F}_0 = \text{circ}\begin{pmatrix} f \\ 0 \\ 0 \\ 0 \end{pmatrix}$$

is a solution of the Dirichlet problem

$$\mathcal{D}^T \mathbf{F} = \mathbf{0}, \quad \text{in } \Omega,$$

$$\mathbf{F} = \mathbf{G}_0, \quad \text{on } \Gamma$$

whence by Theorem 8 we have that $\mathcal{H}[\mathbf{G}_0] = \mathbf{G}_0$. The desired conclusion (ii) then directly follows by comparing the entries in the above equality.

(ii)→ (iii)

From the condition H0,0g=H2,2g=g it follows that C±0,0g=C±2,2g. Therefore, as C0,0g−C2,2g is harmonic in Ω±Ω± and C±0,0g−C±2,2g|Γ=0, we have C0,0g=C2,2g n R4n\Γ. Using the remaining conditions in (ii) and following a similar reasoning as above, we obtain that g satisfies the relations (3) and hence by Theorem 6 we have that C0,0g=C1,1g=C2,2g=C3,3g. Consequently, we obtain that H0,0g=H1,1g=H2,2g=H3,3g=g, as stated in (iii).

(iii)→ (i)

The conditions $H_{0,0}g = H_{1,1}g = H_{2,2}g = H_{3,3}g = g$ imply the solvability of the Dirichlet problems

$$\partial_{\underline{X},r} f = 0, \quad \text{in } \Omega,$$

$$f = g, \quad \text{on } \Gamma, \tag{9}$$

where r=0,...,3. Now, let f0, f1, f2, f3 be the respective solutions of (9), then these functions all are solutions of the classical Dirichlet problem

$$\Delta_{4n} f = 0, \quad \text{in } \Omega,$$

$$f = g, \quad \text{on } \Gamma$$

whence they coincide. The function f=f0=f1=f2=f3 thus is *q*-monogenic and constitutes a solution of (8). □

For right *q*-monogenic functions the following analogue is obtained.

Theorem 10 *Let* g∈C0,α(Γ)$g \in C^{0,\alpha}(\Gamma)$, *then the following statements are equivalent*:

(i)The Dirichlet problem

$$f\partial_{\underline{Z}_0} = f\partial_{\underline{Z}_1} = f\partial_{\underline{Z}_2} = f\partial_{\underline{Z}_3} = 0, \quad in\ \Omega,$$

$$f = g, \quad on\ \Gamma \tag{10}$$

has a solution.

(ii)g satisfies the relations

$$gH_{0,0} = gH_{2,2} = g, \qquad gH_{1,3} = -gH_{3,1}, \qquad gH_{1,1} + gH_{3,3} = 2g.$$

(iii)g satisfies the relations

$$gH_{0,0} = gH_{1,1} = gH_{2,2} = gH_{3,3}g = g.$$

We now turn our attention towards establishing a connection between the two-sided **Q**-monogenicity of a matrix function ***G*** and the matrix Hilbert transforms H[G|Γ] and [G|Γ]H of its trace on the boundary Γ.

Theorem 11 *Let* G∈C0,α(Ω∪Γ), *such that* DTG=0 *in* Ω, *then the following statements are equivalent*:

*(i)**G** is two-sided **Q**-monogenic in* Ω.

*(ii)*H[G|Γ]=[G|Γ]H.

Proof Assume that, next to its already assumed left **Q**-monogenicity, ***G*** also is right **Q**-monogenic in Ω. Then by Theorem 8 it holds that

$$\mathcal{H}[\mathbf{G}|_{\Gamma}] = \mathbf{G}|_{\Gamma} = [\mathbf{G}|_{\Gamma}]\mathcal{H}.$$

Conversely, suppose that H[G|Γ]=[G|Γ]H. By Theorem 4 and its right-handed version, we conclude that the corresponding left and right **Q**-Hermitian Cauchy transform of ***G***, C[G] and [G]C, have the same boundary values on Γ. This fact, together with their harmonicity, implies that

$$\mathcal{C}[\mathbf{G}] = [\mathbf{G}]\mathcal{C}.$$

On the other hand, from the assumed left **Q**-monogenicity of ***G*** we have G=C[G] and hence

$$\mathbf{G} = \mathcal{C}[\mathbf{G}] = [\mathbf{G}]\mathcal{C}$$

which clearly forces ***G*** to be two-sided **Q**-monogenic. □

The following result illustrates the utility of the above theorem when considering *q*-monogenic functions.

Theorem 12 *Let* g∈C0,α(Ω∪Γ) *be left q-monogenic in* Ω, *then the following statements are equivalent*:

(i)g is two-sided q-monogenic in Ω.

(ii)g satisfies the relations

$$H_{0,0}g = gH_{0,0}, \qquad H_{2,2}g = gH_{2,2},$$

$$H_{1,3}g + H_{3,1}g = gH_{1,3} + gH_{3,1},$$

$$H_{1,1}g + H_{3,3}g = gH_{1,1} + gH_{3,3}.$$

(iii)g satisfies the relations

$$H_{0,0}g = gH_{0,0}, \qquad H_{1,1}g = gH_{1,1}, \qquad H_{2,2}g = gH_{2,2}, \qquad H_{3,3}g = gH_{3,3}.$$

Proof (i) ↔ (ii)

From (i) we see that the matrix function $\mathcal{G}_0$ corresponding to g is two-sided **Q**-monogenic in Ω, whence (ii) follows from Theorem 11(ii) applied to $\mathcal{G}_0$. Conversely, (ii) can be rewritten in the matricial form H[G0|Γ]=[G0|Γ]H, from which (i) follows by observing that the two-sided **Q**-monogenicity of $\mathcal{G}_0$ implied by Theorem 11 is equivalent to the q-monogenicity of g.

(i)↔ (iii)

It follows from (i) that g is two-sided monogenic w.r.t. ∂X−−r, r=0,...,3. We may then invoke [[2], Theorem 3.2] in order to conclude that Hr,rg=gHr,r, r=0,...,3. Conversely, suppose that (iii) holds. Each of the conditions Hr,rg=gHr,r, r=0,...,3, implies the two-sided monogenicity of g in Ω w.r.t. ∂X−−r, r=0,...,3, see again [[2], Theorem 3.2], whence g is two-sided q-monogenic in Ω.

ACKNOWLEDGEMENT

Ricardo Abreu-Blaya and Juan Bory-Reyes wish to thank all members of the Department of Mathematical Analysis of Ghent University, where the paper was written, for the invitation and hospitality. They were supported respectively by the Research Council of Ghent University and by the Research Foundation - Flanders (FWO, project 31506208).

COMPETING INTERESTS

The authors declare that they have no competing interests.

AUTHORS' CONTRIBUTIONS

All authors have worked jointly on the manuscript, which is the result of an intensive collaboration. All authors read and approved the final manuscript.

REFERENCES

1. Abreu-Blaya R, Bory-Reyes J, Peña Peña D: Jump problem and removable singularities for monogenic functions. *J. Geom. Anal.* 2007, 17(1):1-13.

2. Abreu-Blaya R, Bory-Reyes J, Delanghe R, Sommen F: Cauchy integral decomposition of multi-vector valued functions on hypersurfaces.*Comput. Methods Funct. Theory* 2005, 5(1):111-135.

3. Abreu-Blaya, R, Bory-Reyes, J, Brackx, F, De Schepper, H, Sommen, F: Cauchy integral formulae in Hermitian quaternionic Clifford analysis. Complex Anal. Oper. Theory (accepted for publication).

4. Abreu-Blaya, R, Bory-Reyes, J, Brackx, F, De Schepper, H, Sommen, F: Matrix Cauchy and Hilbert transforms in Hermitean quaternionic Clifford analysis. Complex Var. Elliptic Equ. (accepted for publication).

5. Brackx F, Delanghe R, Sommen F Research Notes in Mathematics 76. In *Clifford Analysis*. Pitman, Boston; 1982.

6. Brackx F, Bureš J, De Schepper H, Eelbode D, Sommen F, Souček V: Fundaments of Hermitean Clifford analysis - part I: complex structure. *Complex Anal. Oper. Theory* 2007, 1(3):341-365.

7. Brackx F, Bureš J, De Schepper H, Eelbode D, Sommen F, Souček V: Fundaments of Hermitean Clifford analysis - part II: splitting of h -monogenic equations. *Complex Var. Elliptic Equ.* 2007, 52(10-11):1063-1079.

8. Brackx F, De Schepper H, Sommen F: The Hermitean Clifford analysis toolbox. *Adv. Appl. Clifford Algebras* 2008, 18(3-4):451-487.

9. Brackx F, De Knock B, De Schepper H: A matrix Hilbert transform in Hermitean Clifford analysis. *J. Math. Anal. Appl.* 2008, 344: 1068-1078.

10. Damiano A, Eelbode D, Sabadini I: Quaternionic Hermitian spinor systems and compatibility conditions. *Adv. Geom.* 2011, 11: 169-189.

11. Damiano A, Eelbode D, Sabadini I: Algebraic analysis of Hermitian monogenic functions. *C. R. Acad. Sci. Paris, Ser. I* 2008, 346: 139-142.

12. Delanghe R, Sommen F, Souček V: *Clifford Algebra and Spinor-Valued Functions*. Kluwer Academic, Dordrecht; 1992.

13. Eelbode D:A Clifford algebraic framework for sp(m)sp(m)-invariant differential operators. *Adv. Appl. Clifford Algebras* 2007, 17: 635-649.

14. Gilbert J, Murray M: *Clifford Algebras and Dirac Operators in Harmonic Analysis*. Cambridge University Press, Cambridge; 1991.

15. Gürlebeck K, Habetha K, Sprössig W: *Holomorphic Functions in the Plane and n-Dimensional Space*. Birkhäuser, Basel; 2008. Translated from the 2006 German original

16. Gürlebeck K, Sprössig W: *Quaternionic and Clifford Calculus for Physicists and Engineers*. Wiley, Chichester; 1998.

17. Peña-Peña D, Sabadini I, Sommen F: Quaternionic Clifford analysis: the Hermitian setting. *Complex Anal. Oper. Theory* 2007, 1: 97-113.

9

THE MIXED BOUNDARY VALUE PROBLEM FOR THE INHOMOGENEOUS CIMMINO SYSTEM

Liping Wang1 , Zuoliang Xu2 and Yuying Qiao1

1College of Mathematics and Information Science, Hebei Normal University, Shijiazhuang, Hebei province, 050024, P.R. China.

2School of Information, Renmin University of China, Beijing, 100872, P.R. China.

ABSTRACT

In this article, we first propose a kind of mixed boundary value problem for the inhomogeneous Cimmino system, which consists of first order linear partial differential equations in R4R4. Then, by using the one-to-one correspondence between the theory of quaternion valued hyperholomorphic functions and that of Cimmino system's solutions, we transform the problem as stated above into a problem related to the ψ-hyperholomorphic functions in quaternionic analysis. Moreover, we show the boundedness, Hölder continuity, and generalized derivatives of a kind of singular integral operator ψTC2[g]ψTC2[g] related to ψ-hyperholomorphic functions in quaternionic analysis. Lastly, the solution of the mixed boundary value problem for the inhomogeneous Cimmino system is explicitly described.

KEYWORDS

Cimmino system, quaternionic analysis, ψ-hyperholomorphic functions, Cimmino singular, integral operator, mixed boundary value problem

1 INTRODUCTION

The skew field of quaternions $\mathbb{H}$ gives an example of a noncommutative Clifford algebra with minimal dimension. It serves as a very convenient model of general Clifford constructions. Today, quaternionic analysis is regarded as a broadly accepted branch of classical analysis offering a successful generalization of complex analysis. It studies functions defined on domains in R3 or R4 with values in the skew field of real quaternions $\mathbb{H}$. This theory is centered around the concept of ψ-hyperholomorphic functions related to a so-called structural set ψ of H3H3 or H4H4, respectively.

Quaternionic analysis initiated new solution methods for boundary value problems in several research areas of mathematical physics, in particular in planar fluids, quantum field theory, electromagnetic wave equations *etc.* Many scholars and experts have studied some boundary and initial value problems in higher dimensions by using them, such as Gürlebeck, Sprössig, Adler, Alesker, Yang, and so on [1–5].

The Cimmino system (1.1) offers a natural and elegant generalization to the four-dimensional case of that of Cauchy-Riemann. Cimmino, Dragomir and Lanconelli have done a lot of research on it [6, 7]. Recently, Abreu Blaya *et al.* [8] studied the Dirichlet boundary value problem for the inhomogeneous Cimmino system (1.2). We have

$$\begin{cases} \dfrac{\partial f_0}{\partial x_0} - \dfrac{\partial f_1}{\partial x_1} + \dfrac{\partial f_2}{\partial x_2} - \dfrac{\partial f_3}{\partial x_3} = 0, \\ \dfrac{\partial f_0}{\partial x_1} + \dfrac{\partial f_1}{\partial x_0} - \dfrac{\partial f_2}{\partial x_3} - \dfrac{\partial f_3}{\partial x_2} = 0, \\ \dfrac{\partial f_0}{\partial x_2} - \dfrac{\partial f_1}{\partial x_3} - \dfrac{\partial f_2}{\partial x_0} + \dfrac{\partial f_3}{\partial x_1} = 0, \\ \dfrac{\partial f_0}{\partial x_3} + \dfrac{\partial f_1}{\partial x_2} + \dfrac{\partial f_2}{\partial x_1} + \dfrac{\partial f_3}{\partial x_0} = 0, \end{cases} \tag{1.1}$$

where fmfm (m=0,1,2,) are continuously differentiable $\mathbb{R}$-valued functions in Ω⊂R4. The corresponding inhomogeneous Cimmino system is as follows:

$$\begin{cases} \dfrac{\partial f_0}{\partial x_0} - \dfrac{\partial f_1}{\partial x_1} + \dfrac{\partial f_2}{\partial x_2} - \dfrac{\partial f_3}{\partial x_3} = g_0, \\ \dfrac{\partial f_0}{\partial x_1} + \dfrac{\partial f_1}{\partial x_0} - \dfrac{\partial f_2}{\partial x_3} - \dfrac{\partial f_3}{\partial x_2} = g_1, \\ \dfrac{\partial f_0}{\partial x_2} - \dfrac{\partial f_1}{\partial x_3} - \dfrac{\partial f_2}{\partial x_0} + \dfrac{\partial f_3}{\partial x_1} = g_2, \\ \dfrac{\partial f_0}{\partial x_3} + \dfrac{\partial f_1}{\partial x_2} + \dfrac{\partial f_2}{\partial x_1} + \dfrac{\partial f_3}{\partial x_0} = g_3, \end{cases} \tag{1.2}$$

where fmfm are as stated above, gm∈Lp(Ω,R)gm∈Lp(Ω,R) (m=0,1,2,3).

In this article, we will study a kind of mixed boundary value problem for the inhomogeneous Cimmino system (1.2) by using the quaternionic analysis approach. This article is organized as follows. In Section 2, we recall some basic knowledge of quaternionic analysis. In Section 3, we construct a singular integral operator and study some of its properties. In Section 4, we first propose a kind of mixed boundary value problem for the inhomogeneous Cimmino system (1.2); then we obtain an integral representation of the solution of the mixed boundary value problem by using the one-to-one correspondence between the theory of quaternion valued hyperholomorphic functions and that of a Cimmino system's solutions.

2 PRELIMINARIES

Quaternionic analysis studies functions defined on R4R4 with their values in quaternion algebra space $\mathbb{H}$, which is a four-dimensional vector space with basis *e, i, j, k*. The basis element *e* is a unit element, henceforth we shall abbreviate *e* to 1. Also, *i,j, k* satisfy the following multiplication rule:

$$i^2 = j^2 = k^2 = -1, \qquad ij = -ji = k, \qquad jk = -kj = i, \qquad ki = -ik = j.$$

An arbitrary element of the quaternion algebra space $\mathbb{H}$ can be written as $jx_2 + kx_3,\ x_m \in \mathbb{R}\ (m = 0,1,2,3)$, and $\bar{x} = x_0 - ix_1 - jx_2 - kx_3.$, The norm for an element x∈Hx∈H is taken to be $|x| = \sqrt{x_0^2 + x_1^2 + x_2^2 + x_3^2}$ and satisfies $|\bar{x}| = |x|,\ |x+y| \leq |x| + |y|,\ |xy| =$. Obviously, $|x||y|$. Obviously, $\overline{xy} = \bar{y}\bar{x}$ and $x\bar{x} = \bar{x}x = |x|^2.$. In addition, suppose the imaginary unit of $\mathbb{C}$ is identified with the basis element *i* in quaternion algebra space $\mathbb{H}$, then for arbitrary z∈C , we have z=x0+ix1 and its complex conjugate z¯=x0−ix1 ix1. In this way it is easily seen that zj=jz¯.

By means of the mapping $x_0 + ix_1 + jx_2 + kx_3 \rightarrow (x_0 + ix_1) + (x_2 + ix_3)j\ (\rightarrow (x_0, x_1, x_2, x_3))$, one can see $\mathbb{H}$ as $\mathbb{C}^2$ (or $\mathbb{R}^4$). From now on, an arbitrary element ξ∈Hξ∈H can be written as $\xi = z_1 + z_2 j,\ z_1, z_2 \in \mathbb{C}$. From the multiplication rule as stated above, for arbitrary $\xi = z_1 + z_2 j,\ \eta = \varsigma_1 + \varsigma_2 j \in \mathbb{H},\ z_1, z_2, \varsigma_1, \varsigma_2 \in \mathbb{C}.$ we have $\xi\eta = (z_1\varsigma_1 - z_2\bar{\varsigma}_2) + (z_1\varsigma_2 + z_2\bar{\varsigma}_1)j,\ \bar{\xi} = \overline{z_1 + z_2 j} =$ $\bar{z}_1 + \overline{z_2 j} = \bar{z}_1 - z_2 j$, and $\xi\bar{\xi} = \bar{\xi}\xi = |z_1|^2 + |z_2|^2 = |\xi|^2$.

Let Ω⊂R4 be a nonempty open bounded connected set and the boundary Γ=∂Ω be a differentiable, oriented, and compact Liapunov surface. The functions *f* which are defined in Ω with values in $\mathbb{H}$ can be expressed as $f(x) = f_0 + f_1 i + f_2 j + f_3 k$, where $f_m\ (m = 0,1,2,3)$ are continuously differentiable $\mathbb{R}$-valued functions in

$\Omega \subset \mathbb{R}^4$. On $C^{(1)}(\Omega, \mathbb{H})$, we define the differential operators ${}^{\psi}D$ and ψ¯Dψ¯D as follows:

$$ {}^{\psi}D = 2\left(\frac{\partial}{\partial \bar{z}_1} - j\frac{\partial}{\partial \bar{z}_2}\right), \qquad {}^{\bar{\psi}}D = 2\left(\frac{\partial}{\partial z_1} + j\frac{\partial}{\partial \bar{z}_2}\right), $$

where

$$ \frac{\partial}{\partial \bar{z}_1} = \frac{1}{2}\left(\frac{\partial}{\partial x_0} + i\frac{\partial}{\partial x_1}\right), \qquad \frac{\partial}{\partial \bar{z}_2} = \frac{1}{2}\left(\frac{\partial}{\partial x_2} + i\frac{\partial}{\partial x_3}\right), $$

$$ \frac{\partial}{\partial z_1} = \frac{1}{2}\left(\frac{\partial}{\partial x_0} - i\frac{\partial}{\partial x_1}\right), \qquad \frac{\partial}{\partial z_2} = \frac{1}{2}\left(\frac{\partial}{\partial x_2} - i\frac{\partial}{\partial x_3}\right). $$

Obviously, the differential operators ${}^{\psi}D$ and ${}^{\bar{\psi}}D$ can be written as

$$ {}^{\psi}D = \frac{\partial}{\partial x_0} + i\frac{\partial}{\partial x_1} - j\frac{\partial}{\partial x_2} + k\frac{\partial}{\partial x_3}, \qquad {}^{\bar{\psi}}D = \frac{\partial}{\partial x_0} - i\frac{\partial}{\partial x_1} + j\frac{\partial}{\partial x_2} - k\frac{\partial}{\partial x_3}, $$

which are associated to the structural set $\psi = \{1, i, -j, k\}$ and $\bar{\psi} = \{1, -i, j, -k\}$, respectively.

Let ΔR4=∑3m=0∂2xmΔR4=∑m=03∂xm2, then the following equalities hold on C(2)(Ω,H)C(2)(Ω,H):

$$ {}^{\psi}D^{\bar{\psi}}D = {}^{\bar{\psi}}D^{\psi}D = \Delta_{\mathbb{R}^4} \cong \Delta_{\mathbb{C}^2} \cong \Delta_{\mathbb{H}}. $$

Taking into account that the multiplication in $\mathbb{H}$ is noncommutative, the functions f are called left ψ-hyperholomorphic in Ω if ${}^{\psi}D[f](\xi) = 0 \ (\xi \in \Omega)$. The functions g are called right ψ-hyperholomorphic in Ω if Ω if $[g]^{\psi}D(\xi) = 0 \ (\xi \in \Omega)$.

Denote by Θ4Θ4 the fundamental solution of the Laplace operator

$$ \Theta_4(\xi) = -\frac{1}{4\pi^2}\frac{1}{|\xi|^2}, $$

and by KψKψ the fundamental solution of the operator ${}^{\psi}D$:

$$\mathcal{K}_{\psi}(\xi) = {}^{\bar{\psi}}D[\Theta_4] = [\Theta_4]^{\bar{\psi}}D = \frac{1}{2\pi^2}\frac{\xi_{\bar{\psi}}}{|\xi|^4} = \frac{1}{2\pi^2}\frac{\bar{z}_1 + \bar{z}_2 j}{(|z_1|^2 + |z_2|^2)^2}.$$

Then the corresponding Cauchy type integral operator is

$${}^{\psi}K_{\Gamma}[f](\xi) = \int_{\Gamma} \mathcal{K}_{\psi}(\eta - \xi) n_{\psi}(\eta) f(\eta)\, d\Gamma_{\eta} = \int_{\Gamma} \mathcal{K}_{\psi}(\eta - \xi)\, d\sigma_{\eta} f(\eta),$$

and the Teodorescu type integral operator is

$${}^{\psi}T_{\Omega}[f](\xi) = -\int_{\Omega} \mathcal{K}_{\psi}(\eta - \xi) f(\eta)\, d\Omega_{\eta}.$$

In this article, $g(x) \in L_p(\mathbb{C}^2, \mathbb{H})$ means that $g(x) \in L_p(E, \mathbb{H})$, $g_{\sigma}(x) = |x|^{-\sigma} g(\frac{x}{|x|^2}) \in L_p(E, \mathbb{H})$, in which $E = \{\xi \,||\xi| \leq 1\}$, σ is a real number, $\|g\|_{L_p} = \|g\|_{L_p(E)} + \|g_{\sigma}\|_{L_p(E)}, p \geq 1$. The following fundamental statements are widely known to hold and can be found in [1, 9, 10], respectively.

Definition 2.1 Suppose that the functions *f*, *g*, φ are defined in Ω with values in $\mathbb{H}$ $f, g \in L_1(\Omega, \mathbb{H})$. If for arbitrary $\varphi \in C_0^{\infty}(\Omega, \mathbb{H}), f, g$ *g*satisfy

$$\int_{\Omega} [\varphi]^{\psi} D(\xi) g(\xi)\, d\Omega_{\xi} + \int_{\Omega} \varphi(\xi) f(\xi)\, d\Omega_{\xi} = 0,$$

then *f* is called a generalized derivative of the function *g*, denoted by f=ψD[g]f=ψD[g].

Lemma 2.1 ([9]) *If* σ1,σ2>0σ1*, , then we have*

$$|\sigma_1^{\gamma} - \sigma_2^{\gamma}| \leq |\sigma_1 - \sigma_2|^{\gamma}.$$

Lemma 2.2 (Integral form of the quaternionic Stokes formula [1])

Let Ω,Γ=∂ΩΩ,Γ=∂Ω *be as stated above and* $f, g \in C^{(1)}(\Omega, \mathbb{H})$, *then*

$$\int_\Gamma g(\xi) n_\psi(\xi) f(\xi)\, d\Gamma_\xi = \int_\Omega \left([g]^\psi D(\xi) \cdot f(\xi) + g(\xi) \cdot {}^\psi D[f](\xi)\right) d\Omega_\xi.$$

Lemma 2.3 (Borel-Pompeiu quaternionic formula [1]) *Let* Ω,Γ=∂ΩΩ,Γ=∂Ω *be as stated above and* f∈C(1)(Ω,H)f∈C(1)(Ω,H), *then for arbitrary* ξ∈Ωξ∈Ω, *we have*

$$\int_\Gamma \mathcal{K}_\psi(\eta - \xi)\, d\sigma_\eta f(\eta) - \int_\Omega \mathcal{K}_\psi(\eta - \xi)^\psi D[f](\eta)\, d_{\Omega_\eta} = f(\xi)$$

And

$$\int_\Gamma f(\eta)\, d\sigma_\eta \mathcal{K}_\psi(\eta - \xi) - \int_\Omega [f]^\psi D(\eta) \mathcal{K}_\psi(\eta - \xi)\, d_{\Omega_\eta} = f(\xi).$$

Lemma 2.4 (Hadamard lemma [10]) *Suppose* Ω *be as stated above. If* α′α′, β′β′ *satisfy* $\alpha', \beta' < 4, \alpha' + \beta' > 4$, *then for all* $x_1, x_2 \in \mathbb{R}^4$ *and* $x_1 \neq x_2$, *we have*

$$\int_\Omega |t - x_1|^{-\alpha'} |t - x_2|^{-\beta'}\, dt \leq M_0(\alpha', \beta') |x_1 - x_2|^{4-\alpha'-\beta'}.$$

3 SOME USEFUL PROPERTIES OF THE CIMMINO SINGULAR INTEGRAL OPERATOR

By means of the idea as stated above, we suppose

$$\xi = z_1 + z_2 j \in \mathbb{H}, \quad z_1 = x_0 + ix_1, z_2 = x_2 + ix_3 \in \mathbb{C},$$

$$\eta = \varsigma_1 + \varsigma_2 j \in \mathbb{H}, \quad \varsigma_1 = y_0 + iy_1, \varsigma_2 = y_2 + iy_3 \in \mathbb{C},$$

$$f(\xi) = f(z_1, z_2) = u_1(z_1, z_2) + u_2(z_1, z_2) j, \quad u_1 = f_0 + if_1, u_2 = f_2 + if_3 \in \mathbb{C},$$

$$g(\xi) = g(z_1, z_2) = v_1(z_1, z_2) + v_2(z_1, z_2) j, \quad v_1 = g_0 + ig_1, v_2 = g_2 + ig_3 \in \mathbb{C}.$$

Then system (1.1) can be written as

$$(1.1) \iff \begin{cases} \partial_{\bar{z}_1} u_1 + \partial_{z_2} \bar{u}_2 = 0, \\ \partial_{\bar{z}_2} u_1 - \partial_{z_1} \bar{u}_2 = 0 \end{cases} \iff {}^{\psi}D[f] = 0. \tag{3.1}$$

Moreover, if the pair (u1,u2)(u1,u2) of continuously differentiable (up to the second order) complex-valued functions give a solution of system (1.1) then

$$\Delta_{\mathbb{R}^4} u_1 \cong \Delta_{\mathbb{C}^2} u_1 = 4\left(\partial^2_{z_1\bar{z}_1} + \partial^2_{z_2\bar{z}_2}\right) u_1 = 0.$$

A similar observation is valid for u¯2 , so u1u1, u¯2 are complex-valued harmonic functions, *i.e.* the set of solutions of system (1.1) contains all holomorphic functions of two complex variables.

Similarly, system (1.2) can be written as

$$(1.2) \iff \begin{cases} 2\partial_{\bar{z}_1} u_1 + 2\partial_{z_2} \bar{u}_2 = v_1, \\ 2\partial_{\bar{z}_2} u_1 - 2\partial_{z_1} \bar{u}_2 = v_2 \end{cases} \iff {}^{\psi}D[f] = g. \tag{3.2}$$

The generalized Teodorescu type integral operator ψTC2[g] can be written as

$$\begin{aligned} {}^{\psi}T_{\mathbb{C}^2}[g](z_1, z_2) &= {}^{\psi}T_{\mathbb{C}^2}[g](\xi) \\ &= -\int_{\mathbb{C}^2} \mathcal{K}_{\psi}(\eta - \xi) g(\eta)\, d_{\mathbb{C}^2,\eta} \\ &= \frac{1}{2\pi^2} \int_{\mathbb{C}^2} \frac{(\bar{z}_1 - \bar{\varsigma}_1) + (\bar{z}_2 - \bar{\varsigma}_2)j}{(|z_1 - \varsigma_1|^2 + |z_2 - \varsigma_2|^2)^2} g(\varsigma_1, \varsigma_2)\, d_{\mathbb{C}^2 \varsigma_1, \varsigma_2} \\ &= {}^{\psi}T^{(1)}_{\mathbb{C}^2}[g](z_1, z_2) + {}^{\psi}T^{(2)}_{\mathbb{C}^2}[g](z_1, z_2) j, \end{aligned}$$

where the Cimmino singular integral operators ${}^{\psi}T^{(1)}_{\mathbb{C}^2}[g], {}^{\psi}T^{(2)}_{\mathbb{C}^2}[g]$ are as follows:

$$^{\psi}T^{(1)}_{\mathbb{C}^2}[g](z_1,z_2) = \frac{1}{2\pi^2}\int_{\mathbb{C}^2}\frac{(\bar{z}_1-\bar{\varsigma}_1)}{(|z_1-\varsigma_1|^2+|z_2-\varsigma_2|^2)^2}g(\varsigma_1,\varsigma_2)\,d_{\mathbb{C}^2}{}_{\varsigma_1,\varsigma_2},$$

$$^{\psi}T^{(2)}_{\mathbb{C}^2}[g](z_1,z_2) = \frac{1}{2\pi^2}\int_{\mathbb{C}^2}\frac{(\bar{z}_2-\bar{\varsigma}_2)}{(|z_1-\varsigma_1|^2+|z_2-\varsigma_2|^2)^2}g(\varsigma_1,\varsigma_2)\,d_{\mathbb{C}^2}{}_{\varsigma_1,\varsigma_2}.$$

Theorem 3.1 *Let E be as stated above.* *If* $g \in L_p(\mathbb{C}^2,\mathbb{H})$, $4 < p < +\infty$, *then we have*

(1) $|^{\psi}T_{\mathbb{C}^2}[g](\xi)| \le M_1(p)\|g\|_{L_p}$, $\xi \in \mathbb{C}^2 \cong \mathbb{R}^4$,
(2) $^{\psi}T_{\mathbb{C}^2}[g] \in C_\beta(\mathbb{C}^2,\mathbb{H}) \cong C_\beta(\mathbb{R}^4,\mathbb{H})$ $(0 < \beta = 1 - 4/p < 1)$,
(3) $^{\psi}D(^{\psi}T_{\mathbb{C}^2}[g])(\xi) = g(\xi)$, $\xi \in \mathbb{C}^2 \cong \mathbb{R}^4$.

Proof (1) First, we have

$$\begin{aligned}\left|^{\psi}T_{\mathbb{C}^2}[g](\xi)\right| &\le \frac{1}{2\pi^2}\left|\int_E \frac{(\bar{z}_1-\bar{\varsigma}_1)+(\bar{z}_2-\bar{\varsigma}_2)j}{(|z_1-\varsigma_1|^2+|z_2-\varsigma_2|^2)^2}g(\eta)\,dE_\eta\right|\\ &\quad+\frac{1}{2\pi^2}\left|\int_{\mathbb{C}^2-E}\frac{(\bar{z}_1-\bar{\varsigma}_1)+(\bar{z}_2-\bar{\varsigma}_2)j}{(|z_1-\varsigma_1|^2+|z_2-\varsigma_2|^2)^2}g(\eta)\,d_{(\mathbb{C}^2-E)_\eta}\right|\\ &= O_1+O_2.\end{aligned}$$

By the Hölder inequality, we have

$$\begin{aligned}O_1 &= \frac{1}{2\pi^2}\left|\int_E \frac{(\bar{z}_1-\bar{\varsigma}_1)+(\bar{z}_2-\bar{\varsigma}_2)j}{(|z_1-\varsigma_1|^2+|z_2-\varsigma_2|^2)^2}g(\eta)\,dE_\eta\right|\\ &\le \frac{1}{2\pi^2}\int_E\frac{|z_1-\varsigma_1|+|z_2-\varsigma_2|}{|\xi-\eta|^4}|g(\eta)|\,dE_\eta\\ &\le \frac{1}{\pi^2}\int_E\frac{1}{|\xi-\eta|^3}|g(\eta)|\,dE_\eta \le \frac{1}{\pi^2}\|g\|_{L_p}\left(\int_E\frac{1}{|\xi-\eta|^{3q}}\,dE_\eta\right)^{\frac{1}{q}},\end{aligned}$$

where 1/p+1/q=1

When ξ∈E⁻, because of $< p < +\infty$, $1/p + 1/q = 1$, we have $1 < q < 4/3$. Thus $\int_E \frac{1}{|\xi-\eta|^{3q}}\,dE_\eta$ is bounded. Hence we have

$$\left(\int_E \frac{1}{|\xi-\eta|^{3q}}\, dE_\eta\right)^{\frac{1}{q}} \leq J_1.$$

When $\xi \in \mathbb{C}^2 - \bar{E},$ by Lemma 2.1 and the generalized spherical coordinate, we have

$$\left(\int_E \frac{1}{|\xi-\eta|^{3q}}\, dE_\eta\right)^{\frac{1}{q}} \leq J_2\left(\int_{d_0}^{d_0+2} \rho^{3-3q}\, d\rho\right)^{\frac{1}{q}} \leq J_3,$$

where $\rho = |\xi - \eta|, d_0 = d(\xi, \bar{E}).$

Therefore, for $\forall \xi \in \mathbb{C}^2 \cong \mathbb{R}^4,$ we can obtain

$$O_1 \leq M_1'(p)\|g\|_{L_p}, \quad \xi \in \mathbb{C}^2 \cong \mathbb{R}^4,$$

where $M_1'(p) = \max\{J_1/\pi^2, J_3/\pi^2\}.$

For $\eta \in \mathbb{C}^2 - E,$ we suppose that $\eta = \frac{\bar{\eta}'}{|\eta'|^2},$ then we have $|\eta'| \leq 1.$ Thus by $g \in L_p(\mathbb{C}^2, \mathbb{H}),$ similar to the proof as stated above, we have

$$O_2 \leq M_1''(p)\|g\|_{L_p}.$$

Therefore, we obtain

$$\left|{}^{\psi}T_{\mathbb{C}^2}[g](\xi)\right| \leq M_1(p)\|g\|_{L_p}, \quad \xi \in \mathbb{C}^2 \cong \mathbb{R}^4,$$

where $M_1(p) = M_1'(p) + M_1''(p).$

(2) For arbitrary $\xi', \xi'' \in \mathbb{C}^2 \cong \mathbb{R}^4, \xi' \neq \xi'',$ we have

$$\begin{aligned}
&\left|{}^{\psi}T_{\mathbb{C}^2}[g](\xi') - {}^{\psi}T_{\mathbb{C}^2}[g](\xi'')\right| \\
&= \frac{1}{2\pi^2}\left|\int_{\mathbb{C}^2}\left[\frac{(\bar{z}_1' - \bar{\varsigma}_1) + (\bar{z}_2' - \bar{\varsigma}_2)j}{|\xi' - \eta|^4} - \frac{(\bar{z}_1'' - \bar{\varsigma}_1) + (\bar{z}_2'' - \bar{\varsigma}_2)j}{|\xi'' - \eta|^4}\right]g(\eta)\, d_{\mathbb{C}^2{}_\eta}\right| \\
&\leq \frac{1}{2\pi^2}\left|\int_{E}\left[\frac{(\bar{z}_1' - \bar{\varsigma}_1) + (\bar{z}_2' - \bar{\varsigma}_2)j}{|\xi' - \eta|^4} - \frac{(\bar{z}_1'' - \bar{\varsigma}_1) + (\bar{z}_2'' - \bar{\varsigma}_2)j}{|\xi'' - \eta|^4}\right]g(\eta)\, dE_\eta\right| \\
&\quad + \frac{1}{2\pi^2}\left|\int_{\mathbb{C}^2 - E}\left[\frac{(\bar{z}_1' - \bar{\varsigma}_1) + (\bar{z}_2' - \bar{\varsigma}_2)j}{|\xi' - \eta|^4} - \frac{(\bar{z}_1'' - \bar{\varsigma}_1) + (\bar{z}_2'' - \bar{\varsigma}_2)j}{|\xi'' - \eta|^4}\right]g(\eta)\, d_{(\mathbb{C}^2 - E)_\eta}\right| \\
&= O_3 + O_4
\end{aligned}$$

And

$$\begin{aligned}
O_3 &\leq \frac{1}{2\pi^2}\left|\int_{E}\left[\frac{(\bar{z}_1' - \bar{\varsigma}_1)}{|\xi' - \eta|^4} - \frac{(\bar{z}_1'' - \bar{\varsigma}_1)}{|\xi'' - \eta|^4}\right]g(\eta)\, dE_\eta\right| \\
&\quad + \frac{1}{2\pi^2}\left|\int_{E}\left[\frac{(\bar{z}_2' - \bar{\varsigma}_2)j}{|\xi' - \eta|^4} - \frac{(\bar{z}_2'' - \bar{\varsigma}_2)j}{|\xi'' - \eta|^4}\right]g(\eta)\, dE_\eta\right| \\
&= \left|{}^{\psi}T_E^{(1)}[g](z_1', z_2') - {}^{\psi}T_E^{(1)}[g](z_1'', z_2'')\right| + \left|{}^{\psi}T_E^{(2)}[g](z_1', z_2') - {}^{\psi}T_E^{(2)}[g](z_1'', z_2'')\right| \\
&= I_1 + I_2. \qquad (3.3)
\end{aligned}$$

Since

$$\begin{aligned}
&\left|\frac{\bar{z}_1' - \bar{\varsigma}_1}{|\xi' - \eta|^4} - \frac{\bar{z}_1'' - \bar{\varsigma}_1}{|\xi'' - \eta|^4}\right| \\
&= \left|\frac{(\bar{z}_1' - \bar{\varsigma}_1)|\xi'' - \eta|^2(|z_1'' - \varsigma_1|^2 + |z_2'' - \varsigma_2|^2)}{|\xi' - \eta|^4|\xi'' - \eta|^4}\right. \\
&\qquad \left. - \frac{(|z_1' - \varsigma_1|^2 + |z_2' - \varsigma_2|^2)|\xi' - \eta|^2(\bar{z}_1'' - \bar{\varsigma}_1)}{|\xi' - \eta|^4|\xi'' - \eta|^4}\right| \\
&\leq \frac{|(\bar{z}_1' - \bar{\varsigma}_1)|\xi'' - \eta|^2|z_1'' - \varsigma_1|^2 - |z_1' - \varsigma_1|^2|\xi' - \eta|^2(\bar{z}_1'' - \bar{\varsigma}_1)|}{|\xi' - \eta|^4|\xi'' - \eta|^4} \\
&\quad + \frac{|(\bar{z}_1' - \bar{\varsigma}_1)|\xi'' - \eta|^2|z_2'' - \varsigma_2|^2 - |z_2' - \varsigma_2|^2|\xi' - \eta|^2(\bar{z}_1'' - \bar{\varsigma}_1)|}{|\xi' - \eta|^4|\xi'' - \eta|^4} \\
&= \mathcal{K}_1(\xi', \xi'', \eta) + \mathcal{K}_2(\xi', \xi'', \eta). \qquad (3.4)
\end{aligned}$$

Thus

$$I_1 \le \frac{1}{2\pi^2}\int_E \mathcal{K}_1(\xi',\xi'',\eta)\left|g(\eta)\right| dE_\eta + \frac{1}{2\pi^2}\int_E \mathcal{K}_2(\xi',\xi'',\eta)\left|g(\eta)\right| dE_\eta = I_{1_1} + I_{1_2}. \tag{3.5}$$

Again, because of

$$\begin{aligned}
&\mathcal{K}_1(\xi',\xi'',\eta)\\
&= \frac{|(\bar{z}_1' - \bar{\varsigma}_1)|\xi'' - \eta|^2|z_1'' - \varsigma_1|^2 - |z_1' - \varsigma_1|^2|\xi' - \eta|^2(\bar{z}_1'' - \bar{\varsigma}_1)|}{|\xi' - \eta|^4|\xi'' - \eta|^4}\\
&= \frac{|(\bar{z}_1' - \bar{\varsigma}_1)|\xi'' - \eta|^2(z_1'' - \varsigma_1)(\bar{z}_1'' - \bar{\varsigma}_1) - (\bar{z}_1' - \bar{\varsigma}_1)(z_1' - \varsigma_1)|\xi' - \eta|^2(\bar{z}_1'' - \bar{\varsigma}_1)|}{|\xi' - \eta|^4|\xi'' - \eta|^4}\\
&= \frac{|\bar{z}_1' - \bar{\varsigma}_1||\,|\xi'' - \eta|^2(z_1'' - \varsigma_1) - (z_1' - \varsigma_1)|\xi' - \eta|^2|\,|\bar{z}_1'' - \bar{\varsigma}_1|}{|\xi' - \eta|^4|\xi'' - \eta|^4}\\
&\le \frac{|\bar{z}_1' - \bar{\varsigma}_1|[|\xi'' - \eta|^2|z_1'' - z_1'| + ||\xi'' - \eta|^2 - |\xi' - \eta|^2||z_1' - \varsigma_1|]|\bar{z}_1'' - \bar{\varsigma}_1|}{|\xi' - \eta|^4|\xi'' - \eta|^4}\\
&\le \frac{|\bar{z}_1' - \bar{\varsigma}_1|[|\xi'' - \eta|^2|z_1'' - z_1'| + ||\xi'' - \eta|^2 - |\xi' - \eta|^2||z_1' - \varsigma_1|]|\bar{z}_1'' - \bar{\varsigma}_1|}{|\xi' - \eta|^4|\xi'' - \eta|^4}\\
&\le \frac{|\bar{z}_1' - \bar{\varsigma}_1|[|\xi'' - \eta|^2|z_1'' - z_1'| + |\xi'' - \xi'|(|\xi'' - \eta| + |\xi' - \eta|)|z_1' - \varsigma_1|]|\bar{z}_1'' - \bar{\varsigma}_1|}{|\xi' - \eta|^4|\xi'' - \eta|^4}\\
&\le \frac{|\xi' - \eta|[|\xi'' - \eta|^2|\xi'' - \xi'| + |\xi'' - \xi'|(|\xi'' - \eta| + |\xi' - \eta|)|\xi' - \eta|]|\xi'' - \eta|}{|\xi' - \eta|^4|\xi'' - \eta|^4}\\
&= \left|\xi' - \xi''\right|\sum_{l=1}^{3}\frac{1}{|\xi' - \eta|^{4-l}|\xi'' - \eta|^l}.
\end{aligned} \tag{3.6}$$

Thus by (3.5), (3.6), and the Hölder inequality, we have

$$\begin{aligned}
I_{1_1} &\le \frac{1}{2\pi^2}\int_E \sum_{l=1}^{3}\frac{1}{|\xi' - \eta|^{4-l}|\xi'' - \eta|^l}\left|g(\eta)\right| dE_\eta\left|\xi' - \xi''\right|\\
&\le \frac{1}{2\pi^2}\|g\|_{L_p}\left|\xi' - \xi''\right|\sum_{l=1}^{3}\left(\int_E \frac{1}{|\xi' - \eta|^{(4-l)q}|\xi'' - \eta|^{lq}}\,dE_\eta\right)^{\frac{1}{q}}\\
&= \frac{1}{2\pi^2}\|g\|_{L_p}\left|\xi' - \xi''\right|\sum_{l=1}^{3}(I_{1_1}^{(l)})^{\frac{1}{q}}.
\end{aligned} \tag{3.7}$$

Suppose αl=(4−l)q, βl=lq (l=1,2,). By 1<q<4/, we know

$$0 < \alpha_l, \beta_l < 4, \qquad \alpha_l + \beta_l = 4q > 4.$$

Thus, by Lemma 2.4, for l=1,2,3l=1,2,3, we have

$$\begin{aligned} I_{1_1}^{(l)} &= \int_E \frac{1}{|\xi' - \eta|^{(4-l)q} |\xi'' - \eta|^{lq}} dE_\eta \\ &\leq M_0(\alpha_l, \beta_l) \left|\xi' - \xi''\right|^{4-\alpha_l-\beta_l} \\ &= M_0(\alpha_l, \beta_l) \left|\xi' - \xi''\right|^{4-4q}. \end{aligned} \tag{3.8}$$

Thus, by inequalities (3.7) and (3.8), we obtain

$$\begin{aligned} I_{1_1} &\leq \frac{1}{2\pi^2} \|g\|_{L_p} |\xi' - \xi''| \sum_{l=1}^{3} \left(M_0(\alpha_l, \beta_l) \left|\xi' - \xi''\right|^{4-4q}\right)^{\frac{1}{q}} \\ &\leq J_4 \|g\|_{L_p} \left|\xi' - \xi''\right|^{\beta}, \end{aligned} \tag{3.9}$$

where $0 < \beta = 1 + (4 - 4q)/q = 1 - 4/p < 1.$.

Next, we discuss I12I12.

For arbitrary $\xi', \xi'' \in \mathbb{C}^2 \cong \mathbb{R}^4, \xi' \neq \xi''$, we suppose $|\xi' - \xi''| = \delta$ and construct a sphere $B(\xi', 3\delta)$ with the center at ξ′ξ′ and radius 3*δ*. Next we discuss I12I12 in two cases.

(i) If $B(\xi', 3\delta) \cap \bar{E} \neq \emptyset$, then we may suppose $B(\xi', 3\delta) \cap \bar{E} = \Omega_1, \bar{E} - \Omega_1 = \Omega_2.$. Thus we have

$$\begin{aligned} I_{1_2} &= \frac{1}{2\pi^2}\int_E \mathcal{K}_2(\xi',\xi'',\eta)\left|g(\eta)\right| dE_\eta \\ &= \frac{1}{2\pi^2}\int_{\Omega_1} \mathcal{K}_2(\xi',\xi'',\eta)\left|g(\eta)\right| d\Omega_{1\eta} \\ &\quad + \frac{1}{2\pi^2}\int_{\Omega_2} \mathcal{K}_2(\xi',\xi'',\eta)\left|g(\eta)\right| d\Omega_{2\eta} \\ &= I_{1_2}^{(1)} + I_{1_2}^{(2)}. \end{aligned} \tag{3.10}$$

Again, by inequality (3.4), the Hölder inequality, and the use of a local generalized spherical coordinate, we have

$$\begin{aligned} I_{1_2}^{(1)} &\le \frac{1}{2\pi^2}\int_{\Omega_1} \frac{|\bar{z}_1' - \bar{\varsigma}_1||\xi'' - \eta|^2|z_2'' - \varsigma_2|^2}{|\xi' - \eta|^4|\xi'' - \eta|^4}\left|g(\eta)\right| d\Omega_{1\eta} \\ &\quad + \frac{1}{2\pi^2}\int_{\Omega_1} \frac{|z_2' - \varsigma_2|^2|\xi' - \eta|^2|\bar{z}_1'' - \bar{\varsigma}_1|}{|\xi' - \eta|^4|\xi'' - \eta|^4}\left|g(\eta)\right| d\Omega_{1\eta} \\ &\le \frac{1}{2\pi^2}\int_{\Omega_1} \frac{1}{|\xi' - \eta|^3}\left|g(\eta)\right| d\Omega_{1\eta} + \frac{1}{2\pi^2}\int_{\Omega_1} \frac{1}{|\xi'' - \eta|^3}\left|g(\eta)\right| d\Omega_{1\eta} \\ &\le J_5\|g\|_{L_p}\left[\left(\int_{\Omega_1} \frac{1}{|\xi' - \eta|^{3q}}\, d\Omega_{1\eta}\right)^{\frac{1}{q}} + \left(\int_{\Omega_1} \frac{1}{|\xi'' - \eta|^{3q}}\, d\Omega_{1\eta}\right)^{\frac{1}{q}}\right] \\ &\le J_6\|g\|_{L_p}\left[\left(\int_0^{3\delta} \frac{1}{\rho^{3q-3}}\, d\rho\right)^{\frac{1}{q}} + \left(\int_0^{4\delta} \frac{1}{\rho^{3q-3}}\, d\rho\right)^{\frac{1}{q}}\right] \\ &\le J_7\|g\|_{L_p}\delta^{\frac{4-3q}{q}} = J_7\|g\|_{L_p}\left|\xi' - \xi''\right|^{1-\frac{4}{p}} \\ &= J_7\|g\|_{L_p}\left|\xi' - \xi''\right|^{\beta}. \end{aligned} \tag{3.11}$$

In addition,

$$
\begin{aligned}
&\mathcal{K}_2(\xi',\xi'',\eta)\\
&=\frac{|(\bar{z}_1'-\bar{\varsigma}_1)|\xi''-\eta|^2|z_2''-\varsigma_2|^2-|z_2'-\varsigma_2|^2|\xi'-\eta|^2(\bar{z}_1''-\bar{\varsigma}_1)|}{|\xi'-\eta|^4|\xi''-\eta|^4}\\
&=\frac{|(\bar{z}_1''-\bar{\varsigma}_1+\bar{z}_1'-\bar{z}_1'')|\xi''-\eta|^2|z_2''-\varsigma_2|^2-|z_2'-\varsigma_2|^2|\xi'-\eta|^2(\bar{z}_1''-\bar{\varsigma}_1)|}{|\xi'-\eta|^4|\xi''-\eta|^4}\\
&=\left|\frac{(\bar{z}_1''-\bar{\varsigma}_1)|\xi''-\eta|^2|z_2''-\varsigma_2|^2+(\bar{z}_1'-\bar{z}_1'')|\xi''-\eta|^2|z_2''-\varsigma_2|^2}{|\xi'-\eta|^4|\xi''-\eta|^4}\right.\\
&\qquad\left.-\frac{(\bar{z}_1''-\bar{\varsigma}_1)|z_2'-\varsigma_2|^2|\xi'-\eta|^2}{|\xi'-\eta|^4|\xi''-\eta|^4}\right|\\
&\le\frac{|z_1''-\varsigma_1||\,|\xi''-\eta|^2|z_2''-\varsigma_2|^2-|z_2'-\varsigma_2|^2|\xi'-\eta|^2|}{|\xi'-\eta|^4|\xi''-\eta|^4}\\
&\qquad+\frac{|z_1'-z_1''||\xi''-\eta|^2|z_2''-\varsigma_2|^2}{|\xi'-\eta|^4|\xi''-\eta|^4}.
\end{aligned}
$$

Again, because of

$$
\begin{aligned}
&\left|\left|\xi''-\eta\right|^2\left|z_2''-\varsigma_2\right|^2-\left|z_2'-\varsigma_2\right|^2\left|\xi'-\eta\right|^2\right|\\
&\quad=\left|\left|\xi''-\eta\right|^2\left|z_2''-\varsigma_2\right|^2-\left|\xi''-\eta\right|^2\left|z_2'-\varsigma_2\right|^2\right.\\
&\qquad\left.+\left|\xi''-\eta\right|^2\left|z_2'-\varsigma_2\right|^2-\left|z_2'-\varsigma_2\right|^2\left|\xi'-\eta\right|^2\right|\\
&\quad\le\left|\xi''-\eta\right|^2\left|\left|z_2''-\varsigma_2\right|^2-\left|z_2'-\varsigma_2\right|^2\right|+\left|z_2'-\varsigma_2\right|^2\left|\left|\xi''-\eta\right|^2-\left|\xi'-\eta\right|^2\right|\\
&\quad\le\left|\xi''-\eta\right|^2\left|z_2''-z_2'\right|\left(\left|z_2''-\varsigma_2\right|+\left|z_2'-\varsigma_2\right|\right)+\left|z_2'-\varsigma_2\right|^2\left|\xi''-\xi'\right|\left(\left|\xi''-\eta\right|+\left|\xi'-\eta\right|\right)\\
&\quad\le\left|\xi''-\xi'\right|\sum_{m=0}^{3}\left|\xi''-\eta\right|^{3-m}\left|\xi'-\eta\right|^{m}.
\end{aligned}
$$

Thus, we have

$$
\begin{aligned}
&\mathcal{K}_2(\xi',\xi'',\eta)\\
&\quad\le\frac{|\xi''-\eta||\xi''-\xi'|\sum_{m=0}^{3}|\xi''-\eta|^{3-m}|\xi'-\eta|^{m}+|\xi'-\xi''||\xi''-\eta|^2|\xi''-\eta|^2}{|\xi'-\eta|^4|\xi''-\eta|^4}\\
&\quad=|\xi'-\xi''|\left(\frac{2}{|\xi'-\eta|^4}+\sum_{l=1}^{3}\frac{1}{|\xi'-\eta|^{4-l}|\xi''-\eta|^{l}}\right).
\end{aligned}
\tag{3.12}
$$

So by (3.10) and (3.12), we obtain

$$\begin{aligned} I_{1_2}^{(2)} &= \frac{1}{2\pi^2}\int_{\Omega_2} \mathcal{K}_2(\xi',\xi'',\eta)\left|g(\eta)\right| d\Omega_{2\eta} \\ &\le \left|\xi'-\xi''\right|\frac{1}{\pi^2}\int_{\Omega_2}\frac{|g(\eta)|}{|\xi'-\eta|^4}\,d\Omega_{2\eta} \\ &\quad + \left|\xi'-\xi''\right|\frac{1}{2\pi^2}\int_{\Omega_2}\sum_{l=1}^{3}\frac{|g(\eta)|}{|\xi'-\eta|^{4-l}|\xi''-\eta|^l}\,d\Omega_{2\eta} \\ &= I_{1_2}^{(2)'} + I_{1_2}^{(2)''}. \end{aligned} \tag{3.13}$$

First, similar to the method estimating I11I11, we have

$$I_{1_2}^{(2)''} \le J_8\|g\|_{L_p}\left|\xi'-\xi''\right|^{\beta}. \tag{3.14}$$

Second, when $\eta\in\Omega_2,\ |\xi'-\eta|>3\delta,\ |\xi''-\eta|>2\delta.$ Thus we have

$$\begin{aligned} &2\delta \le \left|\xi'-\eta\right| - \left|\xi'-\xi''\right| \le \left|\xi''-\eta\right| \le \left|\xi'-\xi''\right| + \left|\xi'-\eta\right| = \delta + \left|\xi'-\eta\right|, \\ &\delta \le \left|\xi''-\eta\right| - \left|\xi''-\xi'\right| \le \left|\xi'-\eta\right| \le \left|\xi'-\xi''\right| + \left|\xi''-\eta\right| = \delta + \left|\xi''-\eta\right|. \end{aligned} \tag{3.15}$$

So we know

$$\frac{1}{2} \le \frac{|\xi'-\eta|}{|\xi''-\eta|} \le \frac{3}{2}. \tag{3.16}$$

Thus, by (3.16), the Hölder inequality, and Lemma 2.4, we can obtain

$$
\begin{aligned}
I_{1_2}^{(2)'} &= |\xi' - \xi''| \frac{1}{\pi^2} \int_{\Omega_2} \frac{|g(\eta)|}{|\xi' - \eta|^4} d\Omega_{2\eta} \\
&\leq |\xi' - \xi''| \frac{1}{\pi^2} \int_{\Omega_2} \frac{2|g(\eta)|}{|\xi' - \eta|^3 |\xi'' - \eta|} d\Omega_{2\eta} \\
&\leq J_9 \|g\|_{L_p} |\xi' - \xi''|^{1-\frac{4}{p}} \\
&= J_9 \|g\|_{L_p} |\xi' - \xi''|^{\beta}.
\end{aligned}
\tag{3.17}
$$

So, by (3.13), (3.14), and (3.17), we have

$$I_{1_2}^{(2)} \leq J_{10} \|g\|_{L_p} |\xi' - \xi''|^{\beta}. \tag{3.18}$$

Therefore, by (3.10), (3.11), and (3.18), we can obtain

$$I_{1_2} \leq J_{11} \|g\|_{L_p} |\xi' - \xi''|^{\beta}. \tag{3.19}$$

(ii) If B(ξ′,3δ)∩E⁻=∅ then for arbitrary η∈Eη∈E, we have $|\xi' - \eta| > 3\delta,\ |\xi'' - \eta| > 2\delta.$. Thus similar to the method estimating $I_{1_2}^{(2)}$, we have

$$I_{1_2} \leq J_{12} \|g\|_{L_p} |\xi' - \xi''|^{\beta}. \tag{3.20}$$

So, by (3.19) and (3.20), we have

$$I_{1_2} \leq J_{13} \|g\|_{L_p} |\xi' - \xi''|^{\beta}, \quad \xi', \xi'' \in \mathbb{R}^4 \cong \mathbb{C}^2, \tag{3.21}$$

where J13=max{J11,J12}J13=max{J11,J12}.

Thus, to sum up, by (3.5), (3.9), and (3.21), we obtain

$$I_1 \leq J_{14} \|g\|_{L_p} |\xi' - \xi''|^{\beta}, \quad \xi', \xi'' \in \mathbb{R}^4 \cong \mathbb{C}^2. \tag{3.22}$$

Similarly, we have

$$I_2 \leq J_{15}\|g\|_{L_p}\left|\xi' - \xi''\right|^{\beta}, \quad \xi', \xi'' \in \mathbb{R}^4 \cong \mathbb{C}^2. \tag{3.23}$$

So, by (3.3), (3.22), and (3.23), we obtain

$$O_3 \leq M_2'(p)\|g\|_{L_p}\left|\xi' - \xi''\right|^{\beta}, \quad \xi', \xi'' \in \mathbb{R}^4 \cong \mathbb{C}^2, \tag{3.24}$$

where $M_2'(p) = J_{14} + J_{15}$.

For η∈C2−Eη∈C2−E, we suppose that $\eta = \frac{\bar{\eta}'}{|\eta'|^2}$, then we have $|\eta'| \leq 1$. Thus by $g \in L_p(\mathbb{C}^2, \mathbb{H})$, similar to the proof as stated above, we have

$$O_4 \leq M_2''(p)\|g\|_{L_p}\left|\xi' - \xi''\right|^{\beta}, \quad \xi', \xi'' \in \mathbb{R}^4 \cong \mathbb{C}^2.$$

Therefore, for arbitrary $\xi', \xi'' \in \mathbb{C}^2 \cong \mathbb{R}^4, \xi' \neq \xi''$, we obtain

$$\left|{}^{\psi}T_{\mathbb{C}^2}[g](\xi') - {}^{\psi}T_{\mathbb{C}^2}[g](\xi'')\right| \leq M_2(p)\|g\|_{L_p}\left|\xi' - \xi''\right|^{\beta}, \quad \xi', \xi'' \in \mathbb{R}^4 \cong \mathbb{C}^2,$$

where $M_2(p) = M_2'(p) + M_2''(p)$, *i.e.* ${}^{\psi}T_{\mathbb{C}^2}[g] \in C_\beta(\mathbb{C}^2, \mathbb{H}) \cong C_\beta(\mathbb{R}^4, \mathbb{H})$ $(0 < \beta < 1)$.

(3) For arbitrary $\varphi \in C_0^\infty(\mathbb{C}^2, \mathbb{H})$, there exists a bounded closed set $Q \subset \mathbb{C}^2$, such that $\overline{\operatorname{supp}\varphi} \subset\subset Q$. Thus, by $T_{\mathbb{C}^2}[g](\infty) = 0$, Definition 2.1, Lemma 2.3, and the Fubini theorem, we have

$$\begin{aligned}
&\int_{\mathbb{C}^2} [\varphi]^{\psi}D(\xi)^{\psi}T_{\mathbb{C}^2}[g](\xi)\, d_{\mathbb{C}^2_\xi} \\
&= \lim_{d\to\infty} \int_Q [\varphi]^{\psi}D(\xi)^{\psi}T_{\mathbb{C}^2}[g](\xi)\, dQ_\xi \\
&= -\lim_{d\to\infty} \int_Q [\varphi]^{\psi}D(\xi) \int_{\mathbb{C}^2} \mathcal{K}_\psi(\eta-\xi) g(\eta)\, d_{\mathbb{C}^2_\eta}\, dQ_\xi \\
&= \lim_{d\to\infty} \int_{\mathbb{C}^2} \int_Q [\varphi]^{\psi}D(\xi)\mathcal{K}_\psi(\xi-\eta)\, dQ_\xi g(\eta)\, d_{\mathbb{C}^2_\eta} \\
&= \lim_{d\to\infty} \int_{\mathbb{C}^2} \left[\int_{\partial Q} \varphi(\xi)\, d\sigma_\xi \mathcal{K}_\psi(\xi-\eta) - \varphi(\eta) \right] g(\eta)\, d_{\mathbb{C}^2_\eta} \\
&= -\int_{\mathbb{C}^2} \varphi(\eta) g(\eta)\, d_{\mathbb{C}^2_\eta} = -\int_{\mathbb{C}^2} \varphi(\xi) g(\xi)\, d_{\mathbb{C}^2_\xi},
\end{aligned}$$

where $d = \sup_{\xi',\xi''\in Q} |\xi' - \xi''|$. Hence, in the sense of generalized derivatives, we have ${}^{\psi}D({}^{\psi}T_{\mathbb{C}^2}[g])(\xi) = g(\xi)$. □

Remark 3.1

By the process of proof in Theorem 3.1, it is easy to show that ${}^{\psi}T^{(1)}_{\mathbb{C}^2}[g]$, ${}^{\psi}T^{(2)}_{\mathbb{C}^2}[g] \in C_\beta(\mathbb{C}^2, \mathbb{C}) \cong C_\beta(\mathbb{R}^4, \mathbb{C})$ $(0 < \beta < 1)$.

4 INTEGRAL REPRESENTATION OF SOLUTION OF THE MIXED BOUNDARY VALUE PROBLEM FOR THE INHOMOGENEOUS CIMMINO SYSTEM

In this section, let E=E1×E2 be a bounded domain, ∂Em∂Em (m=1,2) be simply closed curves in the zmzm-plane, and ∂Em∈C(1)μ∂Em∈Cμ(1), 0<μ<1. Without loss of generality, we may consider ∂Em={zm||zm|=1} and Em={zm||zm|<1} (m=1,). Denote by E+m Em+, E−mEm− the inner domain and outer domain of ∂Em∂Em, respectively, and $E^{++} = E_1^+ \times E_2^+, E^{+-} = E_1^+ \times E_2^-, E^{-+} = E_1^- \times E_2^+,$ $E^{--} = E_1^- \times E_2^-, \Gamma = \partial E_1 \times \partial E_2.$

Problem P

The mixed boundary value problem for the inhomogeneous Cimmino system (1.2) is to find a function $f(z_1,z_2) = u_1(z_1,z_2) + u_2(z_1,z_2)j$ satisfying the Cimmino system (1.2) and the following boundary condition:

$$u_1^{++}(t_1,t_2) = G_1(t_1,t_2)u_1^{+-}(t_1,t_2) + G_2(t_1,t_2)u_1^{-+}(t_1,t_2) + G_3(t_1,t_2)u_1^{--}(t_1,t_2) + H(t_1,t_2), \quad t = (t_1,t_2) \in \Gamma, \tag{4.1}$$

$$u_2(t_1,t_2) = h(t_1,t_2), \quad t = (t_1,t_2) \in \partial E, \tag{4.2}$$

where $u_1 = f_0 + if_1,\ u_2 = f_2 + if_3,\ z_1 = x_0 + ix_1,\ z_2 = x_2 + ix_3$. $G_1(z_1,z_2),\ G_2(z_1,z_2),\ G_3(z_1,z_2)$ are analytic in E^{+-}, E^{-+}, E^{--} and are continuous in $\bar{E}^{+-}, \bar{E}^{-+}, \bar{E}^{--}$, respectively, which have no zero. We have $G_m(t_1,t_2)\ (m = 1,2,3),\ H(t_1,t_2) \in C_\alpha(\Gamma,\mathbb{C}),\ h(t_1,t_2) \in C_\alpha(\partial E,\mathbb{C})\ (0 < \alpha < 1)$.

Lemma 4.1 *If* $\Psi \in C^{(2)}(E,\mathbb{H}),\ h \in C_\alpha(\partial E,\mathbb{C})\ (0 < \alpha < 1),\ g \in L_p(\mathbb{C}^2,\mathbb{H})\ (\underline{4 < p < +\infty})$, *then the equation* ψD[Ψ]=0ψD[Ψ]=0 *with the boundary condition* $\bar{w}_2|_{\partial E} = \bar{h}(t_1,t_2) - {}^{\psi}T_{\mathbb{C}^2}^{(2)}[g](t_1,t_2)$ *has the solution* Ψ=w1+w2j=w1+jw¯2 *and*

$$\bar{w}_2(\xi) = \int_{\partial E} \left[\bar{h}(t) - \overline{{}^{\psi}T_{\mathbb{C}^2}^{(2)}[g](t)}\right] \frac{\partial}{\partial \nu} G(\xi,t)\, d\partial E_t,$$

$$w_1(\xi) = \Phi(\xi) + w_0(\xi)$$

or

$$\bar{w}_2(z_1,z_2) = \int_{\partial E} \left[\bar{h}(t_1,t_2) - \overline{{}^{\psi}T_{\mathbb{C}^2}^{(2)}[g](t_1,t_2)}\right] \frac{\partial}{\partial \nu} G(z_1,z_2,t_1,t_2)\, d\partial E_{t_1,t_2},$$

$$w_1(z_1,z_2) = \Phi(z_1,z_2) + w_0(z_1,z_2),$$

where ν *is the unit outward normal on* ∂E, $G(\xi,\eta)$ *is the Green's function in* E=E1×E2, $\Phi(z_1,z_2)$ *is an arbitrary analytic function in* E=E1×E2, *and*

$$^{\psi}T^{(2)}_{\mathbb{C}^2}[g](t_1,t_2) = \frac{1}{2\pi^2}\int_{\mathbb{C}^2}\frac{(\bar{t}_2-\bar{\varsigma}_2)}{(|t_1-\varsigma_1|^2+|t_2-\varsigma_2|^2)^2}g(\varsigma_1,\varsigma_2)\,d_{\mathbb{C}^2_{\varsigma_1,\varsigma_2}},$$

$$w_0(z_1,z_2) = \widetilde{T}_{E_1}[-\partial_{z_2}\bar{w}_2] + \widetilde{T}_{E_2}[\Phi_0],$$

$$\widetilde{T}_{E_1}[-\partial_{z_2}\bar{w}_2] = -\frac{1}{\pi}\int_{E_1}\frac{-\partial_{z_2}\bar{w}_2(\varsigma_1,z_2)}{\varsigma_1-z_1}\,dE_{1\varsigma_1},$$

$$\widetilde{T}_{E_2}[\Phi_0] = -\frac{1}{\pi}\int_{E_2}\frac{\Phi_0(z_1,\varsigma_2)}{\varsigma_2-z_2}\,dE_{2\varsigma_2},$$

$$\Phi_0(z_1,z_2) = \frac{1}{2\pi i}\int_{\partial E_1}\frac{\partial_{z_1}\bar{w}_2(\varsigma_1,z_2)}{\varsigma_1-z_1}\,d\partial E_1\varsigma_1.$$

Proof

From Remark 3.1, we know $^{\psi}T^{(2)}_{\mathbb{C}^2}[g] \in C_\beta(\mathbb{C}^2,\mathbb{C}) \cong C_\beta(\mathbb{R}^4,\mathbb{C})\ (0<\beta<1)$. Thus by [9], we have $\bar{h} - \overline{^{\psi}T^{(2)}_{\mathbb{C}^2}[g]} \in C_\mu(\partial E,\mathbb{C})\ (0<\mu=\min\{\alpha,\beta\}<1)$ So we may construct

$$\bar{w}_2(\xi) = \int_{\partial E}\left[\bar{h}(t) - \overline{^{\psi}T^{(2)}_{\mathbb{C}^2}[g](t)}\right]\frac{\partial}{\partial\nu}G(\xi,t)\,d\partial E_t,$$

where ν is the unit outward normal on ∂E, G(ξ,η) is the Green's function in E=E1×E2,and

$$^{\psi}T^{(2)}_{\mathbb{C}^2}[g](t_1,t_2) = \frac{1}{2\pi^2}\int_{\mathbb{C}^2}\frac{(\bar{t}_2-\bar{\varsigma}_2)}{(|t_1-\varsigma_1|^2+|t_2-\varsigma_2|^2)^2}g(\varsigma_1,\varsigma_2)\,d_{\mathbb{C}^2_{\varsigma_1,\varsigma_2}}.$$

Then w¯2(ξ)w¯2(ξ) is a complex-value harmonic function in *E, i.e.* $\Delta_{\mathbb{C}^2}\bar{w}_2 = 4(\partial^2_{z_1\bar{z}_1} + \partial^2_{z_2\bar{z}_2})\bar{w}_2 = 0.$

Hence

$$\partial_{\bar{z}_1}(\partial_{z_1}\bar{w}_2) = -\partial_{\bar{z}_2}(\partial_{z_2}\bar{w}_2). \tag{4.3}$$

Again, by (3.1), we have

$$^{\psi}D[\Psi] = 0 \iff \begin{cases} \partial_{\bar{z}_1} w_1 + \partial_{z_2}\bar{w}_2 = 0, \\ \partial_{\bar{z}_2} w_1 - \partial_{z_1}\bar{w}_2 = 0 \end{cases} \iff \begin{cases} \partial_{\bar{z}_1} w_1 = -\partial_{z_2}\bar{w}_2, \\ \partial_{\bar{z}_2} w_1 = \partial_{z_1}\bar{w}_2. \end{cases} \tag{4.4}$$

By (4.3), we know $-\partial_{z_2}\bar{w}_2, \partial_{z_1}\bar{w}_2$ satisfy the compatibility condition

$$\partial_{\bar{z}_2}(-\partial_{z_2}\bar{w}_2) = \partial_{\bar{z}_1}(\partial_{z_1}\bar{w}_2).$$

Thus by Theorem 7.2.1 of Chapter 7 in [11], the general solution w1(z1,z2) of system (4.4) possesses the form

$$w_1(z_1, z_2) = \Phi(z_1, z_2) + w_0(z_1, z_2),$$

where Φ(z1,z2) is an arbitrary analytic function in E=E1×E2 and

$$w_0(z_1, z_2) = \widetilde{T}_{E_1}[-\partial_{z_2}\bar{w}_2] + \widetilde{T}_{E_2}[\Phi_0],$$

$$\widetilde{T}_{E_1}[-\partial_{z_2}\bar{w}_2] = -\frac{1}{\pi}\int_{E_1} \frac{-\partial_{z_2}\bar{w}_2(\varsigma_1, z_2)}{\varsigma_1 - z_1}\, dE_{1\varsigma_1},$$

$$\widetilde{T}_{E_2}[\Phi_0] = -\frac{1}{\pi}\int_{E_2} \frac{\Phi_0(z_1, \varsigma_2)}{\varsigma_2 - z_2}\, dE_{2\varsigma_2},$$

$$\Phi_0(z_1, z_2) = \frac{1}{2\pi i}\int_{\partial E_1} \frac{\partial_{z_1}\bar{w}_2(\varsigma_1, z_2)}{\varsigma_1 - z_1}\, d\partial E_1 \varsigma_1.$$

□

Lemma 4.2 *Let* GmGm (m=1,2,3), *H*, w0w0, *E etc. be as stated above. Find a sectionally analytic function* Φ(z1,z2) *in* E−+E−+, E−−E−−, *such that* Φ(z1,z2)Φ(z1,z2) *is continuous in* E++E++, E+−E+−, *and satisfies the boundary condition*

$$\Phi^{++}(t_1,t_2) = G_1(t_1,t_2)\Phi^{+-}(t_1,t_2) + G_2(t_1,t_2)\Phi^{-+}(t_1,t_2) + G_3(t_1,t_2)\Phi^{--}(t_1,t_2) + (G_1 + G_2 + G_3 - 1)\big(w_0(t_1,t_2) + {}^{\psi}T^{(1)}_{\mathbb{C}^2}[g](t_1,t_2)\big) + H(t_1,t_2), \quad t = (t_1,t_2) \in \Gamma, \tag{4.5}$$

Where

$${}^{\psi}T^{(1)}_{\mathbb{C}^2}[g](t_1,t_2) = \frac{1}{2\pi^2}\int_{\mathbb{C}^2} \frac{(\bar{t}_1 - \bar{\varsigma}_1)}{(|t_1 - \varsigma_1|^2 + |t_2 - \varsigma_2|^2)^2} g(\varsigma_1,\varsigma_2)\, d_{\mathbb{C}^2_{\varsigma_1,\varsigma_2}}.$$

Then the solution has the form

$$\Phi(z_1,z_2) = \begin{cases} F(z_1,z_2), & z = (z_1,z_2) \in E^{++}, \\ F(z_1,z_2)/G_1(z_1,z_2), & z = (z_1,z_2) \in E^{+-}, \\ F(z_1,z_2)/G_2(z_1,z_2), & z = (z_1,z_2) \in E^{-+}, \\ -F(z_1,z_2)/G_3(z_1,z_2), & z = (z_1,z_2) \in E^{--}, \end{cases} \tag{4.6}$$

Where

$$F(z_1,z_2) = \frac{1}{(2\pi i)^2}\int_{\partial E_1 \times \partial E_2} \frac{\widetilde{H}(\varsigma_1,\varsigma_2)}{(\varsigma_1 - z_1)(\varsigma_2 - z_2)}\, d\partial E_{1\varsigma_1}\, d\partial E_{2\varsigma_2},$$

and $\widetilde{H} = (G_1 + G_2 + G_3 - 1)(w_0 + {}^{\psi}T^{(1)}_{\mathbb{C}^2}[g]) + H.$

Proof From Remark 3.1, we know ${}^{\psi}T^{(1)}_{\mathbb{C}^2}[g] \in C_\beta(\mathbb{C}^2,\mathbb{C}) \cong C_\beta(\mathbb{R}^4,\mathbb{C})\ (0 < \beta < 1)$. Thus by [9], we have $\widetilde{H} = (G_1 + G_2 + G_3 - 1)(w_0 + {}^{\psi}T^{(1)}_{\mathbb{C}^2}[g]) + H \in C_\mu(\Gamma,\mathbb{C})\ (0 < \mu = \min\{\alpha,\beta\} < 1)$. Hence by Theorem 7.1.2 of Chapter 7 in [11], it is not difficult to verify this lemma. □

Theorem 4.1 *Let E, ∂E etc. be as stated above. If* $g \in L_p(\mathbb{C}^2,\mathbb{H})\ (4 < p < +\infty)$, *then the solution of Problem* P *can be expressed as*

$$f(\xi) = \Psi(\xi) + {}^{\psi}T_{\mathbb{C}^2}[g](\xi),$$

where ${}^{\psi}D[\Psi] = 0$ *and*

$$\begin{cases} \Psi(\xi) = w_1(\xi) + w_2(\xi)j = w_1(\xi) + j\bar{w}_2(\xi), \\ w_1(\xi) = \Phi(\xi) + w_0(\xi), \\ \bar{w}_2(\xi) = \int_{\partial E}[\bar{h}(t) - \overline{{}^{\psi}T^{(2)}_{\mathbb{C}^2}[g](t)}]\frac{\partial}{\partial v}G(\xi,t)\,d\partial E_t, \end{cases}$$

$$\begin{aligned} {}^{\psi}T_{\mathbb{C}^2}[g](\xi) = {}^{\psi}T_{\mathbb{C}^2}[g](z_1,z_2) &= \frac{1}{2\pi^2}\int_{\mathbb{C}^2}\frac{(\bar{z}_1 - \bar{\varsigma}_1) + (\bar{z}_2 - \bar{\varsigma}_2)j}{(|z_1 - \varsigma_1|^2 + |z_2 - \varsigma_2|^2)^2}g(\varsigma_1,\varsigma_2)\,d_{\mathbb{C}^2_{\varsigma_1,\varsigma_2}} \\ &= {}^{\psi}T^{(1)}_{\mathbb{C}^2}[g](z_1,z_2) + {}^{\psi}T^{(2)}_{\mathbb{C}^2}[g](z_1,z_2)j, \end{aligned}$$

herein w0, ${}^{\psi}T^{(2)}_{\mathbb{C}^2}[g]$ *are as stated in Lemma* 4.1, Φ, ${}^{\psi}T^{(2)}_{\mathbb{C}^2}[g]$ *are as stated in Lemma* 4.2.

Proof By Theorem 3.1, we know ${}^{\psi}D[{}^{\psi}T_{\mathbb{C}^2}[g]](\xi) = g(\xi)$, thus ${}^{\psi}D[\Psi(\xi) + {}^{\psi}T_{\mathbb{C}^2}[g](\xi)] = g(\xi)$. Hence, by (3.2), we know the general solution of system (1.2) has the form

$$f(\xi) = \Psi(\xi) + {}^{\psi}T_{\mathbb{C}^2}[g](\xi), \tag{4.7}$$

where ${}^{\psi}D[\Psi] = 0$, $\xi = z_1 + z_2 j$, $f(\xi) = f(z_1,z_2) = u_1(z_1,z_2) + u_2(z_1,z_2)j = u_1(z_1,z_2) + j\bar{u}_2(z_1,z_2)$, $\Psi(\xi) = \Psi(z_1,z_2) = w_1(z_1,z_2) + w_2(z_1,z_2)j = w_1(z_1,z_2) + j\bar{w}_2(z_1,z_2)$, and

$$\begin{aligned} {}^{\psi}T_{\mathbb{C}^2}[g](\xi) = {}^{\psi}T_{\mathbb{C}^2}[g](z_1,z_2) &= \frac{1}{2\pi^2}\int_{\mathbb{C}^2}\frac{(\bar{z}_1 - \bar{\varsigma}_1) + (\bar{z}_2 - \bar{\varsigma}_2)j}{(|z_1 - \varsigma_1|^2 + |z_2 - \varsigma_2|^2)^2}g(\varsigma_1,\varsigma_2)\,d_{\mathbb{C}^2_{\varsigma_1,\varsigma_2}} \\ &= {}^{\psi}T^{(1)}_{\mathbb{C}^2}[g](z_1,z_2) + {}^{\psi}T^{(2)}_{\mathbb{C}^2}[g](z_1,z_2)j = {}^{\psi}T^{(1)}_{\mathbb{C}^2}[g](z_1,z_2) + j\overline{{}^{\psi}T^{(2)}_{\mathbb{C}^2}[g]}(z_1,z_2). \end{aligned}$$

Thus

$$\bar{u}_2(z_1,z_2) = \bar{w}_2(z_1,z_2) + \overline{{}^{\psi}T^{(2)}_{\mathbb{C}^2}[g]}(z_1,z_2).$$

So the boundary condition (4.2) in Problem P can be written as

$$\bar{w}_2 = \bar{h}(t_1, t_2) - \overline{{}^{\psi}T^{(2)}_{\mathbb{C}^2}[g]}(t_1, t_2), \quad t = (t_1, t_2) \in \partial E. \tag{4.8}$$

Therefore, by Lemma 4.1, the solution to the equation ψD[Ψ]=0ψD[Ψ]=0 with boundary condition (4.8) can be expressed as

$$\Psi(\xi) = w_1(\xi) + w_2(\xi)j = w_1(\xi) + j\bar{w}_2(\xi),$$

where w1w1, w¯2 are as stated in Lemma 4.1. Again, by (4.7), we have

$$u_1(z_1, z_2) = w_1(z_1, z_2) + {}^{\psi}T^{(1)}_{\mathbb{C}^2}[g](z_1, z_2).$$

From Lemma 4.1, we have

$$w_1(z_1, z_2) = \Phi(z_1, z_2) + w_0(z_1, z_2),$$

where Φ(z1,z2)Φ(z1,z2) is an arbitrary analytic function in E=E1×E2E=E1×E2, w0w0 is as stated in Lemma 4.1. In addition, by Chapter 7 in [11], we know w0∈Cα(C2,C) (0<α<1), by Remark 3.1, we know ${}^{\psi}T^{(1)}_{\mathbb{C}^2}[g] \in C_{\beta}(\mathbb{C}^2, \mathbb{C})\ (0 < \beta < 1)$. So the boundary condition (4.1) in Problem P can be written as

$$\begin{aligned}\Phi^{++}(t_1, t_2) = {} & G_1(t_1, t_2)\Phi^{+-}(t_1, t_2) + G_2(t_1, t_2)\Phi^{-+}(t_1, t_2) \\ & + G_3(t_1, t_2)\Phi^{--}(t_1, t_2) + (G_1 + G_2 + G_3 - 1)\big(w_0(t_1, t_2) \\ & + {}^{\psi}T^{(1)}_{\mathbb{C}^2}[g](t_1, t_2)\big) + H(t_1, t_2), \quad t = (t_1, t_2) \in \Gamma.\end{aligned}$$

Therefore, by Lemma 4.2, we know Φ(z1,z2) can be expressed as (4.6) in Lemma 4.2. In conclusion, we complete the proof. □

ACKNOWLEDGEMENTS

This work was supported by the National Science Foundation of China (No. 11401162, No. 11171349, No. 11301136), the Natural Science Foundation of Hebei Province (No. A2015205012, No. A2014205069, No. A2014208158) and Hebei Normal University Dr. Fund (No. L2015B03, No. L2015B04).

COMPETING INTERESTS

The authors declare that they have no competing interests.

AUTHORS' CONTRIBUTIONS

LPW has presented the main purpose of the article. All authors read and approved the final manuscript.

REFERENCES

1. Gürlebeck, K, Sprössig, W: Quaternionic and Clifford Calculus for Physicists and Engineers. Wiley, New York (1997)
2. Adler, SL: Quaternionic quantum field theory. Commun. Math. Phys. **104**, 611-656 (1986
3. Sprössig, W: Quaternionic operator methods in fluid dynamics. Adv. Appl. Clifford Algebras **18**, 963-978 (2008)
4. Alesker, S: Quaternionic Monge-Ampère equations. J. Geom. Anal. **13**, 205-238 (2003)

5. Yang, PW, Li, D: HλHλ-Regular vector functions and their boundary value problems. Bound. Value Probl. **2012**, 75 (2012)

6. Cimmino, G: Su alcuni sistemi lineari omogenei di equazioni alle derivate parziali del primo ordine. Rend. Semin. Mat. Univ. Padova **12**, 89-113 (1941)

7. Dragomir, S, Lanconelli, E: On first order linear PDE systems all of whose solutions are harmonic functions. Tsukuba J. Math. **30**, 149-170 (2006)

8. Abreu Blaya, R, Bory Reyes, J, Guzmán Adán, A, Schneider, B: Boundary value problems for the Cimmino system via quaternionic analysis. Appl. Math. Comput. **219**, 3872-3881 (2012)

9. Iftimie, V: Functions hypercomplexes. Bull. Math. Soc. Sci. Math. Roum. **57**, 279-332 (1965)

10. Gilbert, RP, Buchanan, JL: First Order Elliptic Systems: A Function Theoretic Approach. Academic Press, New York (1983)

11. Wen, GC: Recent Progress in Theory and Applications of Modern Complex Analysis. Science Press, Beijing (2010)

10

ON COMPLEXIFIED MECHANICS AND COQUATERNIONS

Dorje C Brody and Eva-Maria Graefe

Department of Mathematics, Imperial College London, London SW7 2AZ, UK

ABSTRACT

While real Hamiltonian mechanics and Hermitian quantum mechanics can both be cast in the framework of complex canonical equations, their complex generalizations have hitherto remained tangential. In this communication, quaternionic and coquaternionic (split-signature analogue of quaternions) extensions of Hamiltonian mechanics are introduced and are shown to offer a unifying framework for complexified classical and quantum mechanics. In particular, quantum theories characterized by complex Hamiltonians invariant under spacetime reflection are shown to be equivalent to certain coquaternionic extensions of Hermitian quantum theories. One of the interesting consequences is that the spacetime dimension of these systems is six, not four, on account of the structures of coquaternionic quantum mechanics.

This communication concerns the relation between complexified classical and non-Hermitian quantum mechanics, and their surprising links to quaternionic and coquaternionic mechanics. The main finding is that complexified mechanical systems with real energies studied extensively in the literature over the past decade can alternatively be thought of as certain coquaternionic extensions of the underlying real mechanical systems. This identification leads to the possibility of employing algebraic techniques of quaternions and coquaternions to tackle some of the challenging open issues in complexified classical and quantum mechanics.

Complex (i.e. non-Hermitian) Hamiltonians have long been employed to describe open quantum systems, decay and scattering phenomena [1]. Further, since the realization that complex operators respecting spacetime reflection (PT) symmetry may possess entirely real spectra [2], there have been considerable research interests in examining both physical and mathematical properties of quantum systems described by non-Hermitian Hamiltonians with real spectra. More recently, the interest in these systems has increased notably, in part owing to experimental realizations of the phenomenon of the PT phase transition and other theoretically predicted effects [3, 4].

Complexified classical mechanics has also been studied intensely both in the context of semiclassical calculations and as a classical analogue of non-Hermitian quantum mechanics [5–7]. At classical level, complex-extended mechanics is characterized by a phase space spanned by complex canonical variables. The quantum analogue of a phase space—in the strict sense of a symplectic manifold upon which the dynamics is governed by Hamilton's equations of motion—is the space of pure states, i.e. the space of rays through the origin of the Hilbert space. On this ray space, the wave equation of Schrödinger in Hermitian quantum mechanics can be expressed as Hamilton's equations of motion [8,9]. An apparent puzzle in this context is that the analysis of *complex* extensions of quantum mechanics has thus far been confined to the *real* state space [10–12] (with the exception of [13]). Therefore, the nature of complexification considered so far for quantum mechanics has a rather different characteristic from that for classical mechanics. If classical phase spaces can be extended into the complex domain, it seems paradoxical that quantum state space cannot be extended in an analogous manner. (Quaternionic quantum mechanics [14, 15] can be viewed as representing 'complex' extensions of the underlying real mechanics for real (Hermitian) quantum systems, although this is not the point of view commonly adopted.)

In this communication, we propose a new approach to these apparently different complex generalizations by making use of quaternionic and coquaternionic formulations. The key idea is that phase-space variables consist of canonical conjugate pairs; hence a complexified mechanical system necessarily involves pairs of complex numbers. On the other hand, pairs of complex numbers can usefully be treated as a single quaternion. It is natural therefore to enquire whether complexified mechanics can be represented more concisely in terms of quaternions, inasmuch as real phase-space variables can more conveniently be represented as a single complex number. A deeper motivation for such a study arises from the fact that symmetry properties of quaternions are closely related to symmetries of (Euclidean) spacetime [16], while complexified dynamical systems discussed here are characterized by Hamiltonians invariant under spacetime reflection (PT) symmetry. The appearance of spacetime symmetry in both complexified mechanics and quaternionic algebra suggests that the latter might constitute a useful tool for the investigation of the former. A further motivation comes from the observation that if the parity operator is assumed trace free, then the real dimensionality of the Hilbert space of a PT-symmetric quantum system is a multiple of four, not two [17]. This suggests that quaternions *a priori* are suitable candidates for the characterization of phase-space variables for such theories.

With these observations in mind, we shall explore the possibility that complexified mechanics can alternatively be viewed as a version of quaternionic mechanics. We shall find, perhaps somewhat unexpectedly, that PT-symmetric systems are in fact related to coquaternions, rather than quaternions. In this connection, it is worth remarking that symmetries of coquaternions are related to the Lorentz group, rather than the Euclidean group, in the sense that every rotation in the Minkowski three-space can be expressed in terms of coquaternions. Hence, PT symmetry *a priori* has more in common with Lorentzian spacetime than Euclidean spacetime. It follows that PT-symmetric quantum mechanics, although related, are not equivalent to the traditional quaternionic quantum theories of [14, 15].

Our hope is that the approach introduced here may ultimately help to tackle some open issues in the area of complexified mechanics that have only been answered at most partially. These include, among others, the classical-quantum correspondence of complexified mechanics [18–21], the derivation of statistical mechanics of classical PT-symmetric systems [22], the characterization of combined systems (e.g. entanglement) in PT-symmetric quantum mechanics, and the formulation of a theory of spacetime that is consistent with complex Hamiltonians.

Quaternions and coquaternions. Although properties of quaternions are well known, we find it convenient to briefly summarize here some of the key relations. The algebra of quaternions involves three imaginary units, denoted here by *i*, *j*, *k*, satisfying the relation

$$i^2 = j^2 = k^2 = ijk = -1, \tag{1}$$

carved into the stone of the Brougham Bridge by Hamilton in an act of 'mathematical vandalism' [23]. It follows that quaternions satisfy the following cyclic relation:

$$ij = -ji = k, \quad jk = -kj = i, \quad ki = -ik = j. \tag{2}$$

Hence, matrix representations of the three imaginary units are just the Pauli matrices multiplied by the complex number.

A generic quaternion can be represented in the form $q = q_0 + iq_1 + jq_2 + kq_3$, where $\{q_i\}_{i=0,1,2,3}$ are all real. It is not difficult to see that quaternions that commute with all other quaternions are real, and that the totality of quaternions that commute with a given quaternion forms a subset isomorphic to complex numbers. The conjugate of a quaternion q is given by $\bar{q} = q_0 - iq_1 - jq_2 - kq_3$. It follows that the squared modulus of q is $\bar{q}q = q_0^2 + q_1^2 + q_2^2 + q_3^2$, and that the inverse of q is $q^{-1} = \bar{q}/(\bar{q}q)$. A quaternion can be expressed in the polar form

$$q = |q|e^{\boldsymbol{i}_q \theta_q} = |q|(\cos\theta_q + \boldsymbol{i}_q \sin\theta_q), \tag{3}$$

where $\boldsymbol{i}_q = (iq_1 + jq_2 + kq_3)/|q|$ and $\theta q = \cos^{-1}(q_0/|q|)$.

The symmetry properties of quaternions are thus closely related to the group *SU*(2). The *SU*(1, 1) analogues of quaternions are the coquaternions [24], also known as split quaternions. They satisfy the relation

$$i^2 = -1, \quad j^2 = k^2 = ijk = +1 \tag{4}$$

in place of (1), and the skew-cyclic relation

$$ij = -ji = k, \quad jk = -kj = -i, \quad ki = -ik = j \tag{5}$$

in place of (2). Like quaternions, the conjugate of a coquaternion $q = q_0 + iq_1 + jq_2 + kq_3$ is $\bar{q} = q_0 - iq_1 - jq_2 - kq_3$. It follows that the squared modulus of a coquaternion is indefinite: $\bar{q}q = q_0^2 + q_1^2 - q_2^2 - q_3^2$. Unlike quaternions, a coquaternion need not have an inverse $q^{-1} = \bar{q}/(\bar{q}q)$ if it is null, i.e. if $\bar{q}q = 0$. The polar decomposition of a coquaternion thus involves trigonometric and hyperbolic functions in the usual way. Unlike quaternions, however, coquaternions do not admit finite-dimensional unitary representations.

Complex formulation of real classical and quantum dynamics. Before we proceed to explore the quaternionic or coquaternionic extensions, it is useful to represent the dynamical theory of elementary mechanical systems in terms of complex phase-space variables [25, 26]. We start with a one-dimensional dynamical system, described by real canonical variables *p* and *x*, and introduce the complex conjugate variables $z = (x + ip)/\sqrt{2}$ and $\bar{z} = (x - ip)/\sqrt{2}$. The canonical equation of motion then reads

$$i\frac{dz}{dt} = \frac{\partial H}{\partial \bar{z}}, \tag{6}$$

where $H = H(z, \bar{z})$ denotes the Hamiltonian. In this formulation, it becomes apparent in which way quantum Schrödinger dynamics can be viewed as a special case of Hamiltonian mechanics [25]. Consider a quantum system characterized more generally by a state vector $|z\rangle$, with components z_n. The Schrödinger equation, with a Hermitian Hamiltonian $\hat{H}$, can then be written in the form $i\dot{z}_n = \mathrm{d}H/\mathrm{d}\bar{z}_n$, where $H = \langle z|\hat{H}|z\rangle = \sum_{mn} H_{mn}\bar{z}_n z_m$ is the expectation value of $\hat{H}$. In particular, the Hamiltonian function of a quantum-mechanical system is bilinear in the canonical variables, that is, it is essentially the Hamiltonian of coupled harmonic oscillators.

With this in mind, we shall devote a great part of this communication to the investigation of the bilinear Hamiltonian $H = \bar{z}z$, leading to the equation of motion $i\,\mathrm{d}z/\mathrm{d}t = z$. The trajectories in the z-plane are concentric circles about the origin, traversed with a constant angular velocity, which define a linear critical point called a *centre*. If we perform a Wick rotation, we obtain $\mathrm{d}z/\mathrm{d}t = z$ (or $\mathrm{d}z/\mathrm{d}t = -z$), which defines a linear critical point called a *focus*. The trajectories are rays leaving (or approaching) the origin, traversed with radially increasing velocity. More generally, consider the equation $\mathrm{d}z/\mathrm{d}t = bz$, where b is a complex constant. If both the real and imaginary parts of b are nonzero, then the corresponding linear critical point is a *vortex*, and the orbits consist of concentric spirals leaving (or approaching) the origin traversed with equal radially increasing velocities. Quantum mechanically, all fixed points of unitary dynamics are centres, whereas in a PT-symmetric quantum theory, a pair of centres can turn into a pair of foci or vortices upon symmetry breaking.

Complexified classical and quantum mechanics. Let us now summarize briefly the conventional way in which classical and quantum mechanics are extended into the complex domain. The complexified classical mechanics typically starts from the real canonical equations of motion $\dot{p} = -\partial H/\partial x$ and $\dot{x} = \partial H/\partial p$, and considers the complexification $p = p_0 + ip_1$ and $x = x_0 + ix_1$. In terms of the four real variables (p_0, x_0, p_1, x_1), the Hamiltonian now takes the form $H = H_0(p_0, x_0, p_1, x_1) + iH_1(p_0, x_0, p_1, x_1)$. Assuming that H is analytic and thus satisfies the Cauchy–Riemann conditions $\partial H_0/\partial x_0 = \partial H_1/\partial x_1$, $\partial H_0/\partial x_1 = -\partial H_1/\partial x_0$, $\partial H_0/\partial p_0 = \partial H_1/\partial p_1$, and $\partial H_0/\partial p_1 = -\partial H_1/\partial p_0$, the equations of motion reduce to

$$\dot{p}_0 = -\frac{\partial H_0}{\partial x_0}, \quad \dot{x}_0 = \frac{\partial H_0}{\partial p_0}, \quad \dot{p}_1 = \frac{\partial H_0}{\partial x_1}, \quad \text{and} \quad \dot{x}_1 = -\frac{\partial H_0}{\partial p_1}. \tag{7}$$

In particular, since the total energy $H_0 + iH_1$ is conserved under the Hamiltonian dynamics, if the initial energy is real so that $H_1 = 0$, then it will remain real. Thus, a complexified Hamiltonian system is equivalent to a real two-dimensional system for which the position and momentum are interchanged in the second dimension [5, 6].

The investigations of 'complexified' quantum mechanics, on the other hand, consider complex Hamiltonians of the form $\hat{H} = \hat{H}_0 + i\hat{H}_1$, while leaving the quantum phase space real. The equation of motion is still given by the Schrödinger equation. That is, writing $z_n = (x_n + ip_n)/\sqrt{2}$ for the components of the state vector, we have $i\dot{z}_n = \mathrm{d}H/\mathrm{d}\bar{z}_n$, where the Hamiltonian function is the expectation value of $\hat{H}$, which is now complex. It is then straightforward to verify that equations of motion for the real phase-space variables p_n, x_n are given by

$$\dot{p}_n = -\frac{\partial H_0}{\partial x_n} + \frac{\partial H_1}{\partial p_n} \quad \text{and} \quad \dot{x}_n = \frac{\partial H_0}{\partial p_n} + \frac{\partial H_1}{\partial x_n}. \tag{8}$$

This structure is a combination of a Hamiltonian symplectic flow generated by H_0 and a Hamiltonian gradient flow generated by H_1, and also appears in the semiclassical limit of non-Hermitian quantum dynamics [18].

The dynamics of complexified quantum systems considered in the literature, for the most part, are characterized by (8), and not by (7). This is the sense in which the nature of complexification in classical mechanics has been different from that of quantum mechanics. To see in which way these two formalisms might be unified by a quaternionic or coquaternionic approach, we shall now investigate the special case of a bilinear Hamiltonian, before turning to the more general case.

Quaternionic and coquaternionic oscillators. We again examine a one-dimensional dynamical system with the Hamiltonian $H = \bar{z}z$, but now z is assumed a quaternion. That is, although the dynamical variable is given by $z = (x + ip)/\sqrt{2}$, we let p and x extend into the complex domain according to the prescription $p \to p_0 + jp_1$ and $x \to x_0 + jx_1$. It follows from the algebraic property (2) that $z = (x_0 + ip_0 + jx_1 + kp_1)/\sqrt{2}$. We assume for simplicity a superselection rule [14] that singles out the complex number i for determining the direction of time. Hence, the canonical equation of motion is given by (6) with $H = p^2{}_0 + p_1{}^2 + x^2{}_0 + x_1{}^2$, and we obtain

$$\begin{cases} \dot{p}_0 = -x_0, & \dot{p}_1 = -x_1 \\ \dot{x}_0 = p_0, & \dot{x}_1 = p_1. \end{cases} \tag{9}$$

Thus, a quaternionic oscillator is equivalent to a two-dimensional oscillator. We remark that working with the real variables (p, x), a natural way of extending the theory into quaternionic domain would not have been apparent.

Let us compare this system with a complex PT-symmetric oscillator [7], where we take the Hamiltonian $H = \frac{1}{2}(p^2 + x^2)$ for which equations of motion read $\dot{p} = -x$ and $\dot{x} = p$, and extend the phase-space variables into the complex domain: $p \to p_0 + ip_1$ and $x \to x_0 + ix_1$. The resulting dynamical equations are given by (9). While equations of motion agree, and indeed they share the same fixed-point structure, these two theories are nevertheless distinct because of boundary conditions: the energy $E = p^2{}_0 + p_1{}^2 + x^2{}_0 + x_1{}^2$ is positive definite for quaternionic oscillators, whereas the energy $E = p^2{}_0 - p_1{}^2 + x^2{}_0 - x_1{}^2$ is indefinite for PT-symmetric oscillators.

Next, we consider a coquaternionic oscillator, with the same superselection rule. (It is worth remarking that unlike quaternionic quantum mechanics where one has to make the choice for the direction of time, in coquaternionic quantum mechanics, the imaginary unit 'i' is naturally preferred on account of the fact that $i^2 = -1$, while $j^2 = k^2 = +1$.) In this case, equations of motion remain the same, but owing to the split signature of coquaternions the energy is given by $E = p^2_0 - p_1^2 + x^2_0 - x_1^2$. Hence, we conclude that a complex PT-symmetric oscillator is equivalent to a coquaternionic oscillator. An important point to note here is the fact that in the complex formulation of mechanics, the construction of the Hamiltonian such as $H = \bar{z}z$ implicitly involves the notion of an inner product, which is not evident in Hamilton's formulation involving real variables (p, x). In the case of a complex PT-symmetric oscillator, this inner product is in effect determined by a PT conjugation, which, owing to the fact that the parity operator is trace free, has a split signature equivalent to a coquaternionic norm.

At the level of classical mechanics, the appearance of a PT inner product seems unproblematic because it has no impact on equations of motion. However, if we attempt to formulate statistical mechanics, for instance, then the use of a PT inner product raises a severe obstacle: for any bounded energy, the energy shell in phase space is not compact; hence, an equilibrium microcanonical distribution cannot be defined. At the quantum level, the issue of indefiniteness in a PT inner product has been recognized earlier because inner products are related to probabilities, which have to be non-negative. To remedy this issue, an alternative inner product, based on a CPT conjugation, has been introduced [27, 28]. One possible way forward in the case of a complex classical oscillator therefore is to modify the inner product such that indefinite components pick up an additional -1. Then energy shells for finite energies become compact, allowing, in particular, for a rigorous formulation of statistical mechanics.

Coquaternionic mechanics with real and complex energies. More generally, consider now a real-valued Hamiltonian $H(z, \bar{z})$, where $z = (x_0 + ip_0 + jx_1 + kp_1)/\sqrt{2}$ is a coquaternion, and $\sqrt{2}\,\partial/\partial\bar{z} = \partial/\partial x_0 - i\partial/\partial p_0 + j\partial/\partial x_1 + k\partial/\partial p_1$. From (6), we thus deduce that

$$i\dot{z} = \frac{1}{\sqrt{2}}\left[\frac{\partial H}{\partial x_0} + i\frac{\partial H}{\partial p_0} - j\frac{\partial H}{\partial x_1} - k\frac{\partial H}{\partial p_1}\right], \tag{10}$$

and hence that

$$\dot{p}_0 = -\frac{\partial H}{\partial x_0}, \quad \dot{x}_0 = \frac{\partial H}{\partial p_0}, \quad \dot{p}_1 = \frac{\partial H}{\partial x_1}, \quad \text{and} \quad \dot{x}_1 = -\frac{\partial H}{\partial p_1}. \tag{11}$$

Comparing these equations with (7) where the initial energy is assumed real, we find therefore that a PT-symmetric classical system with real energy is equivalent to a coquaternionic mechanical system with real energy. Note that in a quaternionic mechanical system, the role of p_1 and x_1 are interchanged, and thus it is merely equivalent to a two-dimensional real system.

An alternative case to consider is where the phase space or the state-space variables are kept real so that z remains a complex number, but the energy $H = H_0 + iH_1$ is made complex. In this case, it follows from (6) that equations of motion read

$$\dot{p} = -\frac{\partial H_0}{\partial x} + \frac{\partial H_1}{\partial p} \quad \text{and} \quad \dot{x} = \frac{\partial H_0}{\partial p} + \frac{\partial H_1}{\partial x}, \tag{12}$$

which is of course a special case of (8). If we now allow both energy and phase-space variables be complex, then in the coquaternionic case, we obtain

$$\begin{aligned} \dot{p}_0 &= -\frac{\partial H_0}{\partial x_0} + \frac{\partial H_1}{\partial p_0}, & \dot{x}_0 &= \frac{\partial H_0}{\partial p_0} + \frac{\partial H_1}{\partial x_0}, \\ \dot{p}_1 &= \frac{\partial H_0}{\partial x_1} + \frac{\partial H_1}{\partial p_1}, & \dot{x}_1 &= -\frac{\partial H_0}{\partial p_1} + \frac{\partial H_1}{\partial x_1}. \end{aligned} \tag{13}$$

These equations constitute natural generalizations of (12) in the sense that they embody the structure of a combination of a Hamiltonian symplectic flow generated by H_0 and a Hamiltonian gradient flow generated by H_1.

We remark, more generally, that if the Hamiltonian is coquaternionic valued so that $H = H_0 + iH_1 + jH_2 + kH_3$, then (10) implies

$$\dot{p}_0 = -\frac{\partial H_0}{\partial x_0} + \frac{\partial H_1}{\partial p_0} + \frac{\partial H_2}{\partial x_1} + \frac{\partial H_3}{\partial p_1}, \qquad \dot{x}_0 = \frac{\partial H_0}{\partial p_0} + \frac{\partial H_1}{\partial x_0} - \frac{\partial H_2}{\partial p_1} + \frac{\partial H_3}{\partial x_1},$$
$$\dot{p}_1 = \frac{\partial H_0}{\partial x_1} + \frac{\partial H_1}{\partial p_1} - \frac{\partial H_2}{\partial x_0} + \frac{\partial H_3}{\partial p_0}, \qquad \dot{x}_1 = -\frac{\partial H_0}{\partial p_1} + \frac{\partial H_1}{\partial x_1} + \frac{\partial H_2}{\partial p_0} + \frac{\partial H_3}{\partial x_0} \tag{14}$$

for the dynamical equations.

Two oscillators and two-level systems. The orbit space of a classical system of a pair of oscillators is equivalent mathematically (though not physically) to a quantum two-level system. In particular, the reduced phase space of both systems is just the Bloch sphere S^2. This can be seen by noting that the Hamiltonian of a pair of uncoupled oscillators in complex coordinates is $H = \bar{z}_1 z_1 + \bar{z}_2 z_2$. If the energy is fixed, we obtain the three-sphere $\bar{z}_1 z_1 + \bar{z}_2 z_2 = 1$, upon which the Hamiltonian flow acts as scalar multiplication $z_n \to e^{it} z_n$ [29]. Hence, the orbit space is just the two-sphere resulting from the Hopf map $S^3 \to S^2$.

Similarly, a pair of classical quaternionic oscillators is equivalent to a quaternionic two-level quantum system. In this case, once energy is fixed, we obtain the seven-sphere $\bar{z}_1 z_1 + \bar{z}_2 z_2 = 1$, upon which the Hamiltonian flow acts as quaternionic scalar multiplication. Hence, the associated phase space (Bloch sphere) is a four-sphere S^4 resulting from the Hopf map $S^7 \to S^4$. To see the structure of the state space in the quaternionic case more explicitly, it suffices to note that a generic normalized state vector in a quaternionic two-level system can be written in the parametric form

$$\begin{pmatrix} z_1 \\ z_2 \end{pmatrix} = \begin{pmatrix} \cos\frac{1}{2}\theta \\ \sin\frac{1}{2}\theta\, e^{i_\phi \phi_1} \end{pmatrix}, \tag{15}$$

where $i_\phi = i\cos\phi_2 + j\sin\phi_2\cos\phi_3 + k\sin\phi_2\sin\phi_3$. It is then straightforward to see that $(\theta, \phi_1, \phi_2, \phi_3)$ constitute spherical coordinates for an S^4 in $\mathbb{R}^5$. In higher dimensions, the state space is just the quaternionic projective Hilbert space endowed with a quaternionic Fubini–Study metric arising from transition probabilities. In the case of a pair of coquaternionic oscillators, the resulting orbit space is hyperbolic (a four-dimensional analogue of the Poincaré disc, which is an example of a Siegel domain [30]).

In the case of a quaternionic or coquaternionic two-level system, a generic Hermitian Hamiltonian can be expressed in the form

$$\hat{H} = \tfrac{1}{2}\omega\mathbb{1} + \sum_{l=1}^{5} n_l\hat{\sigma}_l, \tag{16}$$

where $\omega \in \mathbb{R}$, $\vec{n}$ is a unit vector on $S^4 \subset \mathbb{R}^5$, and

$$\hat{\sigma}_1 = \begin{pmatrix} 0 & 1 \\ 1 & 0 \end{pmatrix}, \quad \hat{\sigma}_2 = \begin{pmatrix} 0 & -i \\ i & 0 \end{pmatrix}, \quad \hat{\sigma}_3 = \begin{pmatrix} 1 & 0 \\ 0 & -1 \end{pmatrix}, \\ \hat{\sigma}_4 = \begin{pmatrix} 0 & -j \\ j & 0 \end{pmatrix}, \quad \hat{\sigma}_5 = \begin{pmatrix} 0 & -k \\ k & 0 \end{pmatrix} \tag{17}$$

are the (co)quaternionic Pauli matrices. For a quaternionic Hermitian matrix, the eigenvalues are real, whereas, they are either real or appear as complex conjugate pairs in the case of a coquaternionic Hermitian matrix. In either case, for two-level systems there are six parametric degrees of freedom (consistent with the observation that the most general PT-symmetric 2 × 2 Hamiltonian has six exogenous parameters [31]). In standard quantum mechanics, it follows from the construction of spin–orbit interaction (the so-called Pauli correspondence) that one can identify the spin operators with the three spatial directions. Similarly, (17) suggests that in a quaternionic or coquaternionic quantum mechanics, the ambient space has five dimensions.

It is worth remarking that owing to the lack of commutativity of the imaginary units there are 'observable' effects in quaternionic or coquaternionic mechanics that have no analogue in real mechanics. For instance, a Hamiltonian proportional to an identity matrix can generate nontrivial dynamics [32]. In particular, if we let *i* determine the preferred complex structure, and take the Hamiltonian $\hat{H} = \tfrac{1}{2}\omega\mathbb{1}$, then the dynamics gives rise to a Rabi oscillation between the $\hat{\sigma}_4$ and

$\hat{\sigma}_5$ directions. This can be seen most easily in the Heisenberg picture:

$$\hat{\sigma}_4(t) = \cos(\omega t)\hat{\sigma}_4 + \sin(\omega t)\hat{\sigma}_5, \tag{18}$$

where we have chosen the initial state to be the $\hat{\sigma}_4$ eigenstate. Therefore, we can generate dynamics with a vanishing energy gap. This is reminiscent of the so-called 'arbitrary fast' quantum state transport effect [10–12].

Let us now consider a 2 × 2 coquaternionic Hermitian Hamiltonian and express this in the form

$$\hat{H} = \begin{pmatrix} s+t & q \\ \bar{q} & s-t \end{pmatrix}, \tag{19}$$

where $q = q_0 + iq_1 + jq_2 + kq_3$. The eigenvalues are $E_\pm = s \pm \sqrt{t^2 + \bar{q}q}$, where $\bar{q}q = q_0^2 + q_1^2 - q_2^2 - q_3^2$. Therefore, in the coquaternionic representation, if $t^2 + q_0^2 + q_1^2 > q_2^2 + q_3^2$, then the eigenvalues are real, which can be identified as the region of unbroken PT symmetry; otherwise, the eigenvalues form a complex conjugate pair.

Working in the basis for which $\hat{H}$ is diagonal, the phase-space function for the Hamiltonian is $H = \bar{z}_1 E_+ z_1 + \bar{z}_2 E_- z_2$. (Note that in general $E_\pm$ and z_i need not commute, since $E_\pm$ need not be real.) It follows that equations of motion are given by $\dot{z}_1 = iE_+ z_1$ and $\dot{z}_2 = iE_- z_2$. In particular, if PT symmetry is unbroken, then eigenvalues are real and the fixed points associated with the dynamical evolution are centres, whereas if the symmetry is broken, then the fixed points form a vortex pair. This is a characteristic feature of the so-called PT phase transition whereby a pair of centres coalesce and turn into a pair of vortices (a sink and a source) [33]. Such a transition does not occur in the case of a quaternionic quantum theory, and this establishes the inequivalence of PT-symmetric and quaternionic quantum theories.

Discussion. We have shown that PT-symmetric quantum theories are equivalent to certain coquaternionic extensions of Hermitian quantum theories, at least in finite-dimensional cases. This equivalence suggests that the underlying spacetime dimension is six, not four, on account of the two additional dimensions generated by the additional imaginary units. This follows from the fact that the spin–orbit interaction can only be defined for five spatial dimensions in quaternionic or coquaternionic quantum theories.

Needless to say, there is a substantial literature dealing with various aspects of six-dimensional spacetimes. Apart from the practicality that calculus on a six-dimensional space facilitates various analysis in four dimensions [34–36], models for six-dimensional spacetimes have also been proposed at a fundamental level. Empirical evidences for extra dimensions are often sought at inaccessibly high energies. However, by identifying physical phenomena predicted by coquaternionic quantum theories that cannot be described by standard complex quantum mechanics (such as the Rabi oscillation with a vanishing energy gap described above), it is possible to seek compelling evidences for the existence of extra dimensions at low energies (cf [37] for a discussion on difficulties in detecting quaternionic manifestations in scattering experiments). To this end, it is worth remarking that evidences for octonionic quantum mechanics would suggest that the underlying spacetime has dimension ten.

Some of the challenging open issues of PT-symmetric quantum theories include the formulation of combined systems to characterize such notions as entanglement and coherent states. The mathematical difficulty might be rooted in the fact that quaternionic projective Hilbert spaces (unlike complex projective space of ordinary quantum mechanics [38]) do not admit algebraic varieties that characterize subspaces of particular physical characteristics [39]. Nevertheless, progress has been made in the context of quaternionic quantum mechanics to formulate combined systems [40] or coherent states [41, 42]. It may be that these techniques can be applied to investigate open issues associated with combined systems for PT-symmetric quantum theories.

ACKNOWLEDGMENT

EMG is supported by an Imperial College Junior Research Fellowship.

REFERENCES

1. Moiseyev N 2011 Non-Hermitian Quantum Mechanics (Cambridge: Cambridge University Press)
2. Bender C M and Boettcher S 1998 Real spectra in non-Hermitian Hamiltonians having PT symmetry Phys. Rev. Lett. **80** 5243
3. Guo A et al 2009 Observation of PT-symmetry breaking in complex optical potentials Phys. Rev. Lett. **103** 093902
4. Zhao K F, Schaden M and Wu Z 2010 Enhanced magnetic resonance signal of spin-polarized Rb atoms near surfaces of coated cells Phys. Rev. A **81** 042903
5. Xavier A L Jr and De Aguiar M A M 1996 Complex trajectories in the quartic oscillator and its semiclassical coherent-state propagator Ann. Phys. **252** 458
6. Kaushal R S and Korsch H J 2000 Some remarks on complex Hamiltonian systems Phys. Lett. A**276** 47
7. Bender C M, Boettcher S and Meisinger P N 1999 PT-symmetric quantum mechanics J. Math. Phys. **40** 2201

8. Strocchi F 1966 Complex coordinates and quantum mechanics Rev. Mod. Phys. **38** 36

9. Weinberg S 1989 Testing quantum mechanics Ann. Phys. **194** 336

10. Bender C M, Brody D C, Jones H F and Meister B K 2007 Faster than Hermitian quantum mechanics Phys. Rev. Lett. **98** 040403

11. Mostafazadeh A 2007 Quantum Brachistochrone problem and the geometry of the state space in pseudo Hermitian quantum mechanics Phys. Rev. Lett. **99** 130502

12. Günther U and Samsonov B F 2008 Naimark-dilated PT-symmetric Brachistochrone Phys. Rev. Lett. **101** 230404

13. Nesterov A I 2009 Non-Hermitian quantum systems and time-optimal quantum evolution SIGMA **5**069

14. Finkelstein D, Jaueh J M, Schiminovieh S and Speiser D 1962 Foundations of quaternion quantum mechanics J. Math. Phys. **3** 207

15. Adler S L 1995 Quaternionic Quantum Mechanics and Quantum Fields (Oxford: Oxford University Press)

16. Conway J H and Smith D A 2003 On Quaternions and Octonions (Wellesley, MA: A K Peters)

17. Bender C M, Brody D C, Hughston L P and Meister B K 2007 Geometry of PT-symmetric quantum mechanics arXiv:0704.2959

18. Graefe E M, Höning M and Korsch H J 2010 Classical limit of non-Hermitian quantum dynamics-a generalized canonical structure J. Phys. A: Math. Theor. **43** 075306

19. Bender C M, Hook D W, Meisinger P N and Wang Q 2010 Complex correspondence principle Phys. Rev. Lett. **104** 061601

20. Mostafazadeh A 2006 Real description of classical Hamiltonian dynamics generated by a complex potential Phys. Lett. A **357** 177

21. Curtright T and Mezincescu L 2007 Biorthogonal quantum systems J. Math. Phys. **48** 092106

22. Jones H F and Moreira E S 2010 Quantum and classical statistical mechanics of a class of non-Hermitian Hamiltonians J. Phys. A: Math. Theor. **43** 055307

23. Baez J 2002 The octonions Bull. Am. Math. Soc. **39** 145

24. Cockle J 1849 On systems of algebra involving more than one imaginary; and on equations of the fifth degree Phil. Mag. Series 3 **35** 434

25. Mackey G W 1963 Mathematical Foundations of Quantum Mechanics (New York: Benjamin)

26. Riccia G D and Wiener N 1966 Wave mechanics in classical phase space, Brownian motion, and quantum theory J. Math. Phys. **7** 1372

27. Bender C M, Brody D C and Jones H F 2002 Complex extension of quantum mechanics Phys. Rev. Lett. **89** 270401

28. Mostafazadeh A 2002 Pseudo Hermiticity versus PT-symmetry: III. Equivalence of pseudo-Hermiticity and the presence of antilinear symmetries J. Math. Phys. **43** 3944

29. Atiyah M F 1983 Angular momentum, convex polyhedra, and algebraic geometry Proc. Edinb. Math. Soc. **26** 121

30. Vinberg É B, Gindikin S G and Pjateckiĭ-Šapiro I I 1963 Classification and canonical realisation of complex bounded homogeneous domains Trudy Moskov. Mat. Obsščč **12** 359

31. Wang Q, Chia S and Zhang J 2010 $\mathcal{PT}$ symmetry as a generalisation of Hermiticity J. Phys. A: Math. Theor. **43** 295301

32. Walff U 1981 A quaternion quantum system Phys. Lett. A **84** 89

33. Graefe E M, Korsch H J and Niederle A E 2010 Quantum-classical correspondence for a non-Hermitian Bose-Hubbard dimer Phys. Rev. A **82** 013629

34. Dirac P A M 1936 Wave equations in conformal space Ann. Math. **37** 429

35. Hughston L P and Hurd T R 1983 A CP^5 calculus for space-time fields Phys. Rep. **100** 273

36. Weinberg S 2010 Six-dimensional methods for four-dimensional conformal field theories Phys. Rev. D **82** 045031

37. Adler S L 1988 Scattering and decay theory for quaternionic quantum mechanics, and the structure of induced T nonconservation Phys. Rev. D **37** 3654

38. Brody D C and Graefe E M 2010 Coherent states and rational surfaces J. Phys. A: Math. Theor. **43**255205

39. Berndt J 2003 Symmetric submanifolds of symmetric spaces Proc. 7th Int. Workshop on Differential Geometry (Taegu, 15-16 November 2002) p 1 (available at http://webbuild.knu.ac.kr/~yjsuh/proceedings/7th/%5B1%5D.PDF)

40. Razon A and Horwitz L P 1991 Tensor product of quaternion Hilbert modules Acta Appl. Math. **24**141

41. Adler S L and Millard A C 1997 Coherent states in quaternionic quantum mechanics J. Math. Phys.**38** 2117

42. Thirulogasanthar K, Honnouvo G and Krzyzak A 2010 Coherent states and Hermite polynomials on quaternionic Hilbert spaces J. Phys. A: Math. Theor. **43** 385205

11

A NEW PROOF FOR SOME OPTIMAL INEQUALITIES INVOLVING GENERALIZED NORMALIZED δ-CASORATI CURVATURES

Chul Woo Lee1 , Jae Won Lee2* and Gabriel-Eduard Vîlcu3,4

1Department of Mathematics, Kyungpook National University, Daegu, 702-701, South Korea.

2Department of Mathematics Education, Busan National University of Education, Busan, 611-736, South Korea.

3Faculty of Economic Sciences, Petroleum-Gas University of Ploie¸sti, Bulevardul Bucure¸sti, Nr. 39, Ploie¸sti, 100680, Romania.

4Faculty of Mathematics and Computer Science, Research Center in Geometry, Topology and Algebra, University of Bucharest, Str. Academiei, Nr. 14, Sector 1, Bucure¸sti, 70109, Romania.

ABSTRACT

In this paper we give a new proof for two sharp inequalities involving generalized normalized δ-Casorati curvatures of a slant submanifold in a quaternionic space form. These inequalities were recently obtained in Lee and Vîlcu (Taiwan. J. Math. 19(3):691-702, 2015) using an optimization procedure by showing that a quadratic polynomial in the components of the second fundamental form is parabolic. The new proof is obtained analyzing a suitable constrained extremum problem on submanifold.

KEYWORDS

scalar curvature, mean curvature, δ-Casorati curvature, shape operator, quaternionic space form, slant submanifold, optimal inequality

1. INTRODUCTION

The most powerful tool to find relationships between the main extrinsic invariants and the main intrinsic invariants of a submanifold is provided by Chen's invariants [1]. This theory was initiated in [2]: Chen established a sharp inequality for a submanifold in a real space form using the scalar curvature and the sectional curvature (both being intrinsic invariants) and squared mean curvature (the main extrinsic invariant). On the other hand, it is well known that the Casorati curvature of a submanifold in a Riemannian manifold is an extrinsic invariant defined as the normalized square of the length of the second fundamental form and it was preferred by Casorati over the traditional Gauss curvature because corresponds better with the common intuition of curvature [3–5]. Some optimal Chen-like inequalities involving Casorati curvatures were proved in [6–10] for several submanifolds in real, complex and quaternionic space forms. Recently, two sharp inequalities involving generalized normalized δ-Casorati curvatures of slant submanifolds in quaternionic space forms were obtained in [11] as follows.

Theorem 1.1

Let MnMn *be a θ-slant proper submanifold of a quaternionic space form* $\overline{M}^{4m}(c)$. *Then*:

(i)The generalized normalized δ-Casorati curvature δC(r;n−1) δC(r;n−1) *satisfies*

$$\rho \leq \frac{\delta_C(r;n-1)}{n(n-1)} + \frac{c}{4}\left(1 + \frac{9}{n-1}\cos^2\theta\right) \quad (1)$$

(1) for any real number r such that 0<r<n(n−1)0<r<n(n−1).

(ii)The generalized normalized δ-Casorati curvature δ^C(r;n−1)δ^C (r;n−1) *satisfies*

$$\rho \leq \frac{\widehat{\delta}_C(r;n-1)}{n(n-1)} + \frac{c}{4}\left(1 + \frac{9}{n-1}\cos^2\theta\right) \quad (2)$$

for any real number r>n(n−1).

Moreover, the equality sign holds in the inequalities (1) *and* (2) *if and only if* MnMn *is an invariantly quasi-umbilical submanifold with trivial normal connection in* M¯4m(c)M¯4m(c), *such that with respect to suitable orthonormal tangent frame* {ξ1,...,ξn*and normal orthonormal frame* {ξn+1,...,ξ4m}, *the shape operators* Ar≡Aξr r∈{n+1,...,4m}r∈{n+1,...,4m}, *take the following forms*:

$$A_{n+1} = \begin{pmatrix} a & 0 & 0 & \cdots & 0 & 0 \\ 0 & a & 0 & \cdots & 0 & 0 \\ 0 & 0 & a & \cdots & 0 & 0 \\ \vdots & \vdots & \vdots & \ddots & \vdots & \vdots \\ 0 & 0 & 0 & \cdots & a & 0 \\ 0 & 0 & 0 & \cdots & 0 & \frac{n(n-1)}{r}a \end{pmatrix}, \qquad A_{n+2} = \cdots = A_{4m} = 0. \quad (3)$$

The proof given in [11] for these inequalities is based on an optimization procedure by showing that a quadratic polynomial in the components of the second fundamental form is parabolic. In this work we give an alternative proof using Oprea's optimization method on submanifolds [12], namely analyzing a suitable constrained extremum problem (see also [13–17]).

2. PRELIMINARIES

This section gives several basic definitions and notations for our framework based mainly on [18, 19].

Let MnMn be an n-dimensional Riemannian submanifold of a Riemannian manifold $(\overline{M},\overline{g})$. Then we denote by g the metric tensor induced on M. Let K(π)K(π) be the sectional curvature of M associated with a plane section $\pi \subset T_pM,\ p \in M$. If $\{e_1,\ldots,e_n\}$ is an orthonormal basis of the tangent space TpMTpM and {en+1,...,em}{en+1,...,em} is an orthonormal basis of the normal space T⊥pMTp⊥M, then the scalar curvature τ at p is given by

$$\tau(p) = \sum_{1 \le i < j \le n} K(e_i \wedge e_j)$$

and the normalized scalar curvature ρ of M is defined as

$$\rho = \frac{2\tau}{n(n-1)}.$$

We denote by H the mean curvature vector, that is,

$$H(p) = \frac{1}{n}\sum_{i=1}^{n} h(e_i, e_i),$$

and we also set

$$h_{ij}^{\alpha} = g(h(e_i, e_j), e_\alpha), \quad i, j \in \{1, \ldots, n\}, \alpha \in \{n+1, \ldots, m\}.$$

Then it is well known that the squared mean curvature of the submanifold *M* in $\bar{M}$ is defined by

$$\|H\|^2 = \frac{1}{n^2} \sum_{\alpha=n+1}^{m} \left(\sum_{i=1}^{n} h_{ii}^{\alpha} \right)^2,$$

and the squared norm of *h* over dimension *n* is denoted by CC and is called the Casorati curvature of the submanifold *M*. Therefore we have

$$C = \frac{1}{n} \sum_{\alpha=n+1}^{m} \sum_{i,j=1}^{n} \left(h_{ij}^{\alpha} \right)^2.$$

The submanifold *M* is called *invariantly quasi-umbilical* if there exist m−nm−n mutually orthogonal unit normal vectors ξn+1,...,ξm msuch that the shape operators with respect to all directions ξαξα have an eigenvalue of multiplicity n−1n−1 and that for each ξαξα the distinguished eigendirection is the same.

Suppose now that *L* is an *s*-dimensional subspace of TpM s≥2, and let {e1,...,es}{e1,...,es} be an orthonormal basis of *L*. Then the scalar curvature τ(L)τ(L) of the *s*-plane section *L* is given by

$$\tau(L) = \sum_{1 \le \alpha < \beta \le s} K(e_\alpha \wedge e_\beta)$$

and the Casorati curvature C(L)C(L) of the subspace *L* is defined as

$$C(L) = \frac{1}{s} \sum_{\alpha=n+1}^{m} \sum_{i,j=1}^{s} \left(h_{ij}^{\alpha} \right)^2.$$

The generalized normalized δ-Casorati curvatures $\delta_C(r;n-1)$ and $\widehat{\delta}_C(r;n-1)$ and δˆC(r;n−1)δ^C(r;n−1) of the submanifold MnMn are defined for any positive real number r≠n(n−1)r≠n(n−1) as

$$\left[\delta_C(r;n-1)\right]_p = r\mathcal{C}_p + \frac{(n-1)(n+r)(n^2-n-r)}{rn}\inf\left\{\mathcal{C}(L)|L \text{ a hyperplane of } T_pM\right\},$$

if $0 < r < n^2 - n$, and

$$\left[\widehat{\delta}_C(r;n-1)\right]_p = r\mathcal{C}_p - \frac{(n-1)(n+r)(r-n^2+n)}{rn}\sup\left\{\mathcal{C}(L)|L \text{ a hyperplane of } T_pM\right\},$$

if $r > n^2 - n$.

If ∇ is the Levi-Civita connection on $\bar{M}$ and ∇ is the covariant differentiation induced on M, then the Gauss and Weingarten formulas are given by

$$\overline{\nabla}_X Y = \nabla_X Y + h(X,Y), \quad \forall X, Y \in \Gamma(TM)$$

and

$$\overline{\nabla}_X N = -A_N X + \nabla_X^{\perp} N, \quad \forall X \in \Gamma(TM), \forall N \in \Gamma\left(TM^{\perp}\right),$$

where h is the second fundamental form of M, ∇⊥∇⊥ is the connection on the normal bundle, and ANAN is the shape operator of $M$with respect to N. If we denote by $\bar{R}$ and R the curvature tensor fields of ∇ and ∇, then we have the Gauss equation:

$$\overline{R}(X,Y,Z,W) = R(X,Y,Z,W) + \overline{g}\left(h(X,W),h(Y,Z)\right) - \overline{g}\left(h(X,Z),h(Y,W)\right), \tag{4}$$

for all $X, Y, Z, W \in \Gamma(TM)$.

Assume now that $(\overline{M},\overline{g})$is a smooth manifold such that there is a rank 3-subbundle σ of End$(T\overline{M})$ with local basis {J1,J2,J3}{J1,J2,J3} satisfying for all α∈{1,2,3}α∈{1,2,3}:

$$\overline{g}(J_\alpha\cdot,J_\alpha\cdot)=\overline{g}(\cdot,\cdot)$$

and

$$J_\alpha^2=-\mathrm{Id},\qquad J_\alpha J_{\alpha+1}=-J_{\alpha+1}J_\alpha=J_{\alpha+2},$$

where Id denotes the identity tensor field of type (1,1)(1,1) on $\bar{M}$ and the indices are taken from {1,2,3}{1,2,3} modulo 3. Then $(\overline{M},\sigma,\overline{g})$ is said to be an almost quaternionic Hermitian manifold. It is easy to see that such a manifold is of dimension 4*m*, m≥1m≥1. Moreover, if the bundle σ is parallel with respect to the Levi-Civita connection ∇ of $\bar{g}$, then $(\overline{M},\sigma,\overline{g})$ is said to be a quaternionic Kähler manifold.

Let $(\overline{M},\sigma,\overline{g})$ be a quaternionic Kähler manifold and let X be a non-null vector field on $\bar{M}$. Then the 4-plane spanned by {X,J1X,J2X,J3X}, denoted by Q(X), is called a quaternionic 4-plane. Any 2-plane in Q(X) is called a quaternionic plane. The sectional curvature of a quaternionic plane is called a quaternionic sectional curvature. A quaternionic Kähler manifold is a quaternionic space form if its quaternionic sectional curvatures are equal to a constant, say *c*. It is well known that a quaternionic Kähler manifold $(\overline{M},\sigma,\overline{g})$is a quaternionic space form, denoted $\overline{M}(c)$, if and only if its curvature tensor is given by

$$\overline{R}(X,Y)Z=\frac{c}{4}\Big\{\overline{g}(Z,Y)X-\overline{g}(X,Z)Y+\sum_{\alpha=1}^{3}\big[\overline{g}(Z,J_\alpha Y)J_\alpha X-$$
$$-\overline{g}(Z,J_\alpha X)J_\alpha Y+2\overline{g}(X,J_\alpha Y)J_\alpha Z\big]\Big\} \tag{5}$$

for all vector fields *X*, *Y*, *Z* on $\bar{M}$ and any local basis {J1,J2,J3} of σ.

A submanifold M of a quaternionic Kähler manifold $(\overline{M},\sigma,\overline{g})$ is said to be a slant submanifold [20] if, for each non-zero vector Xtangent to M at p, the angle θ(X) between Jα(X) and TpM α∈{1,2,3} is constant, *i.e.* it does not depend on the choice of p∈Mp∈M and X∈TpM. We can easily see that quaternionic submanifolds are slant submanifolds with θ=0θ=0 and totally real submanifolds are slant submanifolds with θ=π2. A slant submanifold of a quaternionic Kähler manifold is said to be proper (or θ-slant proper) if it is neither quaternionic nor totally real. We recall that every proper slant submanifold of a quaternionic Kähler manifold is of even dimension n=2s≥2 and we can choose a canonical orthonormal local frame, called an adapted slant frame, as follows:

$$\{e_1, e_2 = \sec\theta P_\alpha e_1, \ldots, e_{2s-1}, e_{2s} = \sec\theta P_\alpha e_{2s-1}\},$$

where Pαe2k−1Pαe2k−1 denotes the tangential component of Jαe2k−1Jαe2k−1, k∈{1,...,s}, and α is 1, 2 or 3 (see [21]).

Let $(\overline{M},\sigma,\overline{g})$ be a Riemannian manifold, M be a submanifold of $\bar{M}$, g be the induced metric of $\bar{g}$ and f:M⟶Rf: be a differentiable function. If we consider the constrained extremum problem

$$\min_{x\in M} f(x), \tag{6}$$

then we have the following result.

Lemma 2.1

[12]

If x0∈Mx0∈M *is the solution of the problem* (6), *then*

(i $(\mathrm{grad}(f))(x_0) \in T^{\perp}_{x_0}M;$

(ii)the bilinear form

$$\mathcal{A}: T_{x_0}M \times T_{x_0}M \longrightarrow \mathbb{R},$$

$$\mathcal{A}(X,Y) = \mathrm{Hess}_f(X,Y) + \overline{g}\big(h(X,Y), (\mathrm{grad}(f))(x_0)\big)$$

is positive semi-definite, where h is the second fundamental form of M in $\bar{M}$ *and* grad(f)grad(f) *is the gradient of f.*

3 NEW PROOF OF THEOREM *1.1*

Since M is θ-slant, it is well known from [20] that

$$P_\beta P_\alpha X = -\cos^2\theta X, \quad \forall X \in \Gamma(TM), \alpha, \beta \in \{1,2,3\}, \tag{7}$$

where PαXPαX denotes the tangential component of JαXJαX.

From (7) it follows immediately that

$$g(P_\alpha X, P_\beta Y) = \cos^2\theta g(X,Y) \tag{8}$$

for X,Y∈Γ(TM) and α,β∈{1,2,3}.

On the other hand, because $\overline{M}^{4m}(c)$ is a quaternionic space form, from (4) and (5) we derive

$$n^2\|H\|^2 = 2\tau(p) + \|h\|^2 - \frac{n(n-1)c}{4} - \frac{3c}{4}\sum_{\beta=1}^{3}\sum_{i,j=1}^{n} g^2(P_\beta e_i, e_j). \tag{9}$$

Choosing now an adapted slant basis

$$\{e_1, e_2 = \sec\theta P_\alpha e_1, \ldots, e_{2s-1}, e_{2s} = \sec\theta P_\alpha e_{2s-1}\}$$

of $T_pM, p \in M$, where $2s = n$, from (7) and (8), we derive

$$g^2(P_\beta e_i, e_{i+1}) = g^2(P_\beta e_{i+1}, e_i) = \cos^2\theta, \quad \text{for } i = 1, 3, \ldots, 2s-1, \tag{10}$$

and

$$g(P_\beta e_i, e_j) = 0, \quad \text{for } (i,j) \notin \{(2l-1, 2l), (2l, 2l-1) | l \in \{1, 2, \ldots, s\}\}. \tag{11}$$

By using (10) and (11) in (9) we get

$$2\tau(p) = n^2\|H\|^2 - nC + \frac{c}{4}\left[n(n-1) + 9n\cos^2\theta\right]. \tag{12}$$

We consider now the following quadratic polynomial in the components of the second fundamental form:

$$\mathcal{P} = rC + \frac{(n-1)(n+r)(n^2-n-r)}{rn}C(L) - 2\tau(p) + \frac{c}{4}\left[n(n-1) + 9n\cos^2\theta\right],$$

where L is a hyperplane of TpMTpM. Without loss of generality we can assume that L is spanned by e1,...,en−1e1,...,en−1. Then we derive

$$\begin{aligned}\mathcal{P} = {} & \frac{r}{n}\sum_{\alpha=n+1}^{4m}\sum_{i,j=1}^{n}\left(h_{ij}^{\alpha}\right)^2 + \frac{(n+r)(n^2-n-r)}{rn}\sum_{\alpha=n+1}^{4m}\sum_{i,j=1}^{n-1}\left(h_{ij}^{\alpha}\right)^2 \\ & - 2\tau(p) + \frac{c}{4}\left[n(n-1) + 9n\cos^2\theta\right].\end{aligned} \tag{13}$$

From (12) and (13), we obtain

$$\mathcal{P} = \sum_{\alpha=n+1}^{4m} \sum_{i=1}^{n-1} \left[\frac{n^2 + n(r-1) - 2r}{r} \left(h_{ii}^{\alpha}\right)^2 + \frac{2(n+r)}{n} \left(h_{in}^{\alpha}\right)^2 \right]$$
$$+ \sum_{\alpha=n+1}^{4m} \left[\frac{2(n+r)(n-1)}{r} \sum_{i<j=1}^{n-1} \left(h_{ij}^{\alpha}\right)^2 - 2 \sum_{i<j=1}^{n} h_{ii}^{\alpha} h_{jj}^{\alpha} + \frac{r}{n} \left(h_{nn}^{\alpha}\right)^2 \right]$$
$$\geq \sum_{\alpha=n+1}^{4m} \left[\sum_{i=1}^{n-1} \frac{n^2 + n(r-1) - 2r}{r} \left(h_{ii}^{\alpha}\right)^2 - 2 \sum_{i<j=1}^{n} h_{ii}^{\alpha} h_{jj}^{\alpha} + \frac{r}{n} \left(h_{nn}^{\alpha}\right)^2 \right]. \tag{14}$$

For r=n+1,…,4m, let us consider the quadratic form fα:Rn⟶Rfα:Rn⟶R defined by

$$f_{\alpha}\left(h_{11}^{\alpha}, \ldots, h_{nn}^{\alpha}\right) = \sum_{i=1}^{n-1} \frac{n^2 + n(r-1) - 2r}{r} \left(h_{ii}^{\alpha}\right)^2 - 2 \sum_{i<j=1}^{n} h_{ii}^{\alpha} h_{jj}^{\alpha} + \frac{r}{n} \left(h_{nn}^{\alpha}\right)^2,$$

and the constrained extremum problem

$\min f_{\alpha}$

subject to $F : h_{11}^{\alpha} + \cdots + h_{nn}^{\alpha} = c^{\alpha}$,

where cαcα is a real constant.

The partial derivatives of fαfα are

$$\begin{cases} \frac{\partial f_{\alpha}}{\partial h_{ii}^{\alpha}} = \frac{2(n+r)(n-1)}{r} h_{ii}^{\alpha} - 2 \sum_{k=1}^{n} h_{kk}^{\alpha} = 0, \\ \frac{\partial f_{\alpha}}{\partial h_{nn}^{\alpha}} = \frac{2r}{n} h_{nn}^{\alpha} - 2 \sum_{k=1}^{n-1} h_{kk}^{\alpha} = 0, \end{cases} \tag{15}$$

with i∈{1,…,n−1}I, i≠j and α∈{n+1,…,4m}

For an optimal solution (hα11,…,hαnn)(h11α,…,hnnα) of the problem, the vector grad(fα)grad(fα) is normal at *F*. That is, it is collinear with the vector (1,1,...,1)(1,1,...,1). From (15), it follows that a critical point of the corresponding problem has the form

$$\begin{cases} h_{ii}^{\alpha} = \frac{r}{n(n-1)} t^{\alpha}, & i \in \{1,\ldots,n-1\}, \\ h_{nn}^{\alpha} = t^{\alpha}. \end{cases} \tag{16}$$

Using (16) and $\sum_{i=1}^{n} h_{ii}^{\alpha} = c^{\alpha}$, we derive

$$\begin{cases} h_{ii}^{\alpha} = \frac{r}{(n+r)(n-1)} c^{\alpha}, & i \in \{1,\ldots,n-1\}, \\ h_{nn}^{\alpha} = \frac{n}{n+r} c^{\alpha}. \end{cases} \tag{17}$$

We fix an arbitrary point x∈Fx∈F. The 2-form A:TxF×TxF⟶RA:TxF×TxF⟶R has the form

$$\mathcal{A}(X,Y) = \mathrm{Hess}(f_{\alpha})(X,Y) + \langle h(X,Y), (\mathrm{grad}(f))(x_0)\rangle,$$

where h is the second fundamental form of F in RnRn and ⟨,⟩⟨,⟩ is the standard inner product on RnRn. Moreover, it is easy to see that the Hessian matrix of fαfα has the form

$$\mathrm{Hess}(f_{\alpha}) = \begin{pmatrix} \frac{2(n+r)(n-1)}{r} - 2 & -2 & \cdots & -2 & -2 \\ -2 & \frac{2(n+r)(n-1)}{r} - 2 & \cdots & -2 & -2 \\ \vdots & \vdots & \ddots & \vdots & \vdots \\ -2 & -2 & \cdots & \frac{2(n+r)(n-1)}{r} - 2 & -2 \\ -2 & -2 & \cdots & -2 & \frac{2r}{n} \end{pmatrix}.$$

As F is totally geodesic in RnRn, considering a vector $X = (X_1, X_2, \ldots, X_n)$ tangent to F at the arbitrary point x on F (that is, verifying the relation $\sum_{i=1}^{n} X_i = 0$), we obtain

$$\begin{aligned}\mathcal{A}(X,X) &= \frac{2(n^2-n+rn-2r)}{r}\sum_{i=1}^{n-1}X_i^2+\frac{2r}{n}X_n^2-2\left(\sum_{i=1}^{n}X_i\right)^2\\ &= \frac{2(n^2-n+rn-2r)}{r}\sum_{i=1}^{n-1}X_i^2+\frac{2r}{n}X_n^2\\ &\geq 0.\end{aligned}$$

Hence the point $(h_{11}^{\alpha},\ldots,h_{nn}^{\alpha})$ from (16) is a global minimum point by Lemma 2.1. Moreover, fα$(h_{11}^{\alpha},\ldots,h_{nn}^{\alpha})=0.$. Therefore, we have

$$\mathcal{P}\geq 0, \tag{18}$$

and this implies

$$2\tau(p)\leq rC+\frac{(n-1)(n+r)(n^2-n-r)}{rn}C(L)+\frac{c}{4}\left[n(n-1)+9n\cos^2\theta\right].$$

Therefore we derive

$$\rho\leq\frac{r}{n(n-1)}C+\frac{(n+r)(n^2-n-r)}{rn^2}C(L)+\frac{c}{4}\left(1+\frac{9}{n-1}\cos^2\theta\right) \tag{19}$$

for every tangent hyperplane L of M and both inequalities (1) and (2) obviously follow from (19).

Moreover, we can easily see now that the equality sign holds in the inequalities (1) and (2) if and only if

$$h_{ij}^{\alpha}=0,\quad \forall i,j\in\{1,\ldots,n\}, i\neq j, \tag{20}$$

and

$$h_{nn}^{\alpha}=\frac{n(n-1)}{r}h_{11}^{\alpha}=\frac{n(n-1)}{r}h_{22}^{\alpha}=\cdots=\frac{n(n-1)}{r}h_{n-1,n-1}^{\alpha} \tag{21}$$

for all α∈{n+1,...,4m}α∈{n+1,...,4m}.

Finally, from (20) and (21) we deduce that the equality sign holds in (1) and (2) if and only if the submanifold M is invariantly quasi-umbilical with trivial normal connection in $\bar{M}$, such that the shape operators take the forms (3) with respect to suitable tangent and normal orthonormal frames.

ACKNOWLEDGEMENTS

The third author was supported by CNCS-UEFISCDI, project number PN-II-ID-PCE-2011-3-0118.

COMPETING INTERESTS

The authors declare that they have no competing interests.

AUTHORS' CONTRIBUTIONS

All authors contributed equally to the writing of this paper. All authors read and approved the final manuscript.

REFERENCES

1. Chen, B-Y: Mean curvature and shape operator of isometric immersions in real-space-forms. Glasg. Math. J. **38**(1), 87-97 (1996)

2. Chen, B-Y: Some pinching and classification theorems for minimal submanifolds. Arch. Math. **60**, 568-578 (1993)

3. Casorati, F: Mesure de la courbure des surfaces suivant l'idée commune. Ses rapports avec les mesures de courbure gaussienne et moyenne. Acta Math. **14**(1), 95-110 (1890)

4. Kowalczyk, D: Casorati curvatures. Bull. Transilv. Univ. Braşov Ser. III **1**(50), 209-213 (2008)

5. Verstraelen, L: Geometry of submanifolds I. The first Casorati curvature indicatrices. Kragujev. J. Math. **37**(1), 5-23 (2013)

6. Decu, S, Haesen, S, Verstraelen, L: Optimal inequalities involving Casorati curvatures. Bull. Transilv. Univ. Brasov Ser. B (N.S.) **14**(49), suppl., 85-93 (2007)

7. Decu, S, Haesen, S, Verstraelen, L: Optimal inequalities characterising quasi-umbilical submanifolds. J. Inequal. Pure Appl. Math. **9**(3), 79 (2008)

8. Ghişoiu, V: Inequalities for the Casorati curvatures of slant submanifolds in complex space forms. In: Riemannian Geometry and Applications. Proceedings RIGA 2011, pp. 145-150. Ed. Univ. Bucureşti, Bucharest (2011)

9. Lee, CW, Yoon, DW, Lee, JW: Optimal inequalities for the Casorati curvatures of submanifolds of real space forms endowed with semi-symmetric metric connections. J. Inequal. Appl. **2014**, Article ID 327 (2014)

10. Slesar, V, Şahin, B, Vîlcu, GE: Inequalities for the Casorati curvatures of slant submanifolds in quaternionic space forms. J. Inequal. Appl.**2014**, Article ID 123 (2014)

11. Lee, JW, Vîlcu, GE: Inequalities for generalized normalized δ-Casorati curvatures of slant submanifolds in quaternionic space forms. Taiwan. J. Math. **19**(3), 691-702 (2015)

12. Oprea, T: Optimization methods on Riemannian submanifolds. An. Univ. Bucur., Mat. **54**(1), 127-136 (2005)

13. Malek, F, Nejadakbary, V: A lower bound for the Ricci curvature of submanifolds in generalized Sasakian space forms. Adv. Geom. **13**(4), 695-711 (2013)

14. Oprea, T: Ricci curvature of Lagrangian submanifolds in complex space forms. Math. Inequal. Appl. **13**(4), 851-858 (2010)

15. Oprea, T: Chen's inequality in the Lagrangian case. Colloq. Math. **108**(1), 163-169 (2007)

16. Rapcsák, T: Sectional curvatures in nonlinear optimization. J. Glob. Optim. **40**(1-3), 375-388 (2008)

17. Zhang, P: Inequalities for Casorati curvatures of submanifolds in real space forms. arXiv:1408.4996

18. Chen, B-Y: Pseudo-Riemannian Geometry, δ-Invariants and Applications. World Scientific, Hackensack (2011)

19. Ishihara, S: Quaternion Kählerian manifolds. J. Differ. Geom. **9**, 483-500 (1974)

20. Şahin, B: Slant submanifolds of quaternion Kaehler manifolds. Commun. Korean Math. Soc. **22**(1), 123-135 (2007)

21. Vîlcu, GE, Chen, B-Y: Inequalities for slant submanifolds in quaternionic space forms. Turk. J. Math. **34**(1), 115-128 (2010)

INDEX